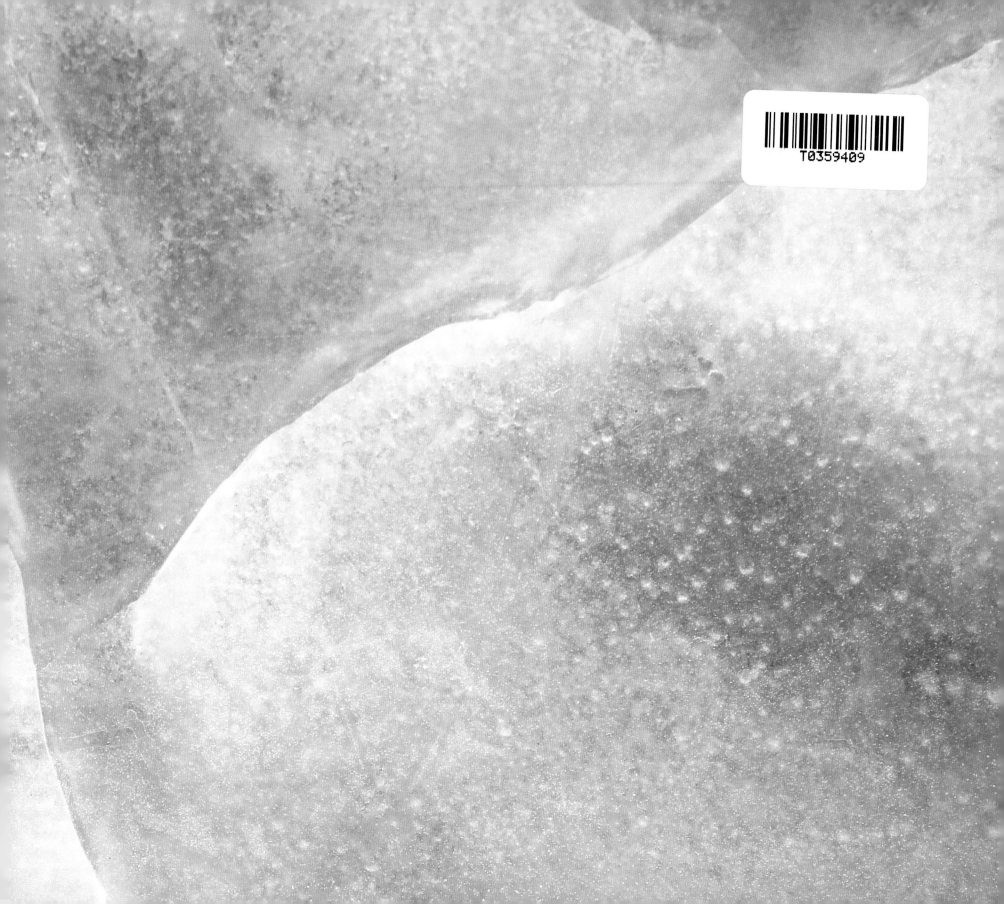

ANTARCTICA

ANTARCTICA

NINA GALLO

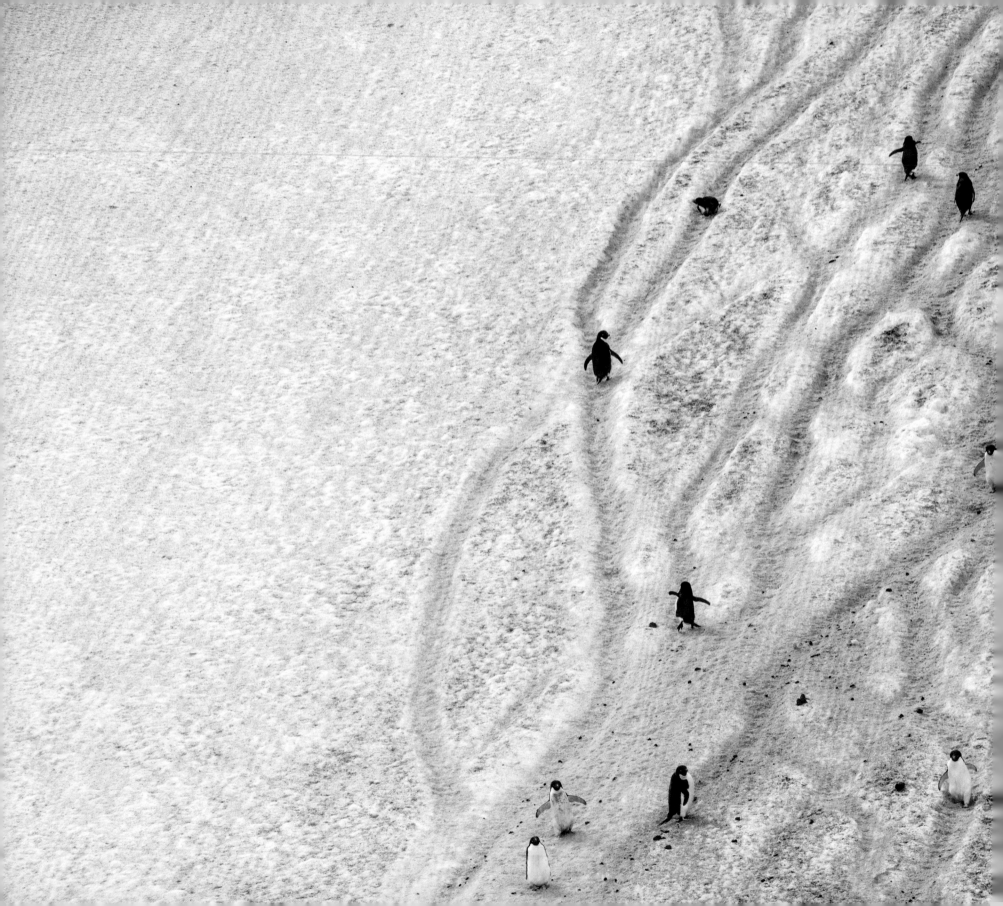

CONTENTS

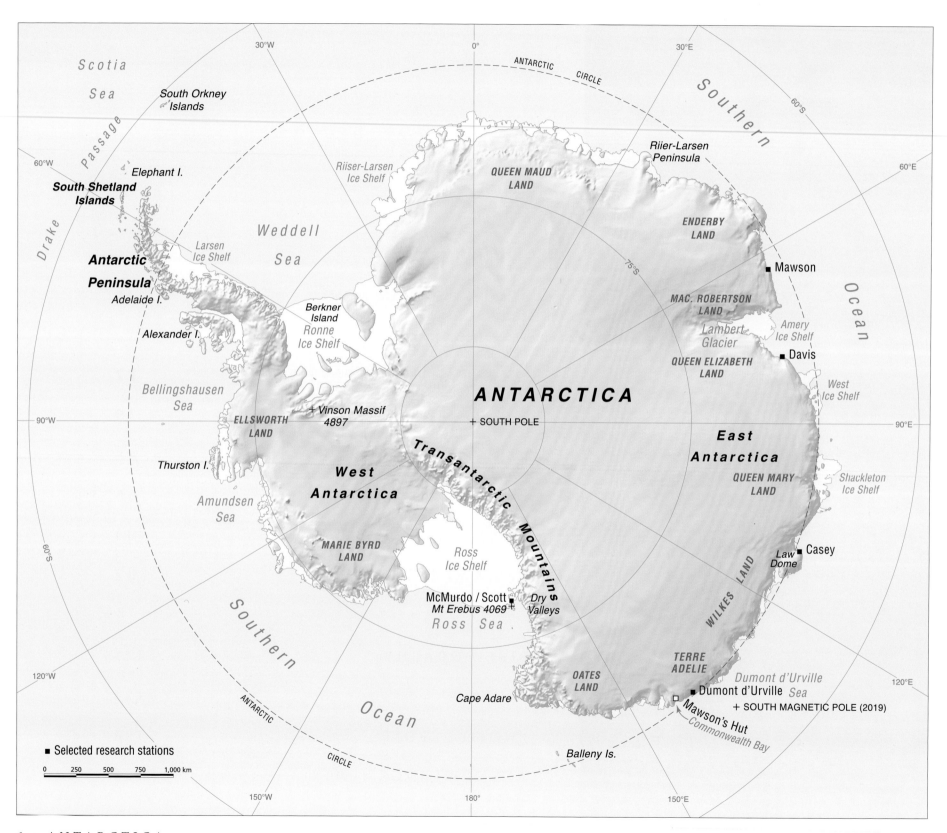

Scotia Sea

South Orkney Islands

Drake Passage

Elephant I.

South Shetland Islands

Antarctic Peninsula

Adelaide I.

Alexander I.

Bellingshausen Sea

ELLSWORTH LAND

Thurston I.

Amundsen Sea

West Antarctica

MARIE BYRD LAND

Southern Ocean

Weddell Sea

Larsen Ice Shelf

Berkner Island

Ronne Ice Shelf

+ Vinson Massif 4897

Ross Ice Shelf

McMurdo / Scott
Mt Erebus 4069 +
Ross Sea

Dry Valleys

Riiser-Larsen Ice Shelf

ANTARCTIC   CIRCLE

30°W   0°   30°E

60°W

60°S

90°W

60°S

120°W

ANTARCTIC   CIRCLE

150°W   180°   150°E

QUEEN MAUD LAND

Riier-Larsen Peninsula

Southern

60°E

ENDERBY LAND

75°S

■ Mawson

MAC. ROBERTSON LAND

Lambert Glacier

Amery Ice Shelf

Ocean

QUEEN ELIZABETH LAND

■ Davis

West Ice Shelf

90°E

ANTARCTICA

+ SOUTH POLE

Transantarctic Mountains

East Antarctica

QUEEN MARY LAND

Shackleton Ice Shelf

WILKES LAND

Law Dome

■ Casey

TERRE ADELIE

OATES LAND

Cape Adare

Balleny Is.

Dumont d'Urville Sea

□ Dumont d'Urville

+ SOUTH MAGNETIC POLE (2019)

Mawson's Hut
Commonwealth Bay

120°E

■ Selected research stations

0   250   500   750   1,000 km

INTRODUCTION

IN JANUARY 1820, after millennia spent speculating and searching, humans laid eyes on Antarctica for the very first time. Since then, our relationship with this enigmatic land of ice and snow has continually evolved, from one of industry during the sealing and whaling eras, to exploration as pioneers charted undiscovered coastlines and forged perilous paths towards the South Pole.

Today, the human relationship with Antarctica is defined by groundbreaking scientific endeavour and extraordinary international collaboration. As the 21st century progresses, the significance of this remote continent is more tangible than ever. The Antarctic tourism industry has burgeoned in recent years, welcoming tens of thousands of visitors into a rapidly changing environment. Current scientific research is bringing into focus the startling impacts that climate change in Antarctica could have on a global scale. Both tourism and research were severely impacted by the global COVID-19 pandemic of 2020 and the future of cruise-based tourism remains uncertain. Climate research has never been more important.

The bicentenary of the first human sighting of Antarctica is an opportunity to reflect on our Antarctic legacy and consider how the next 50 years will shape the Antarctic age of our time. Discover this remarkable continent covered in ice, where days endure for months, highways are built by penguins and shimmering lights illuminate the night sky.

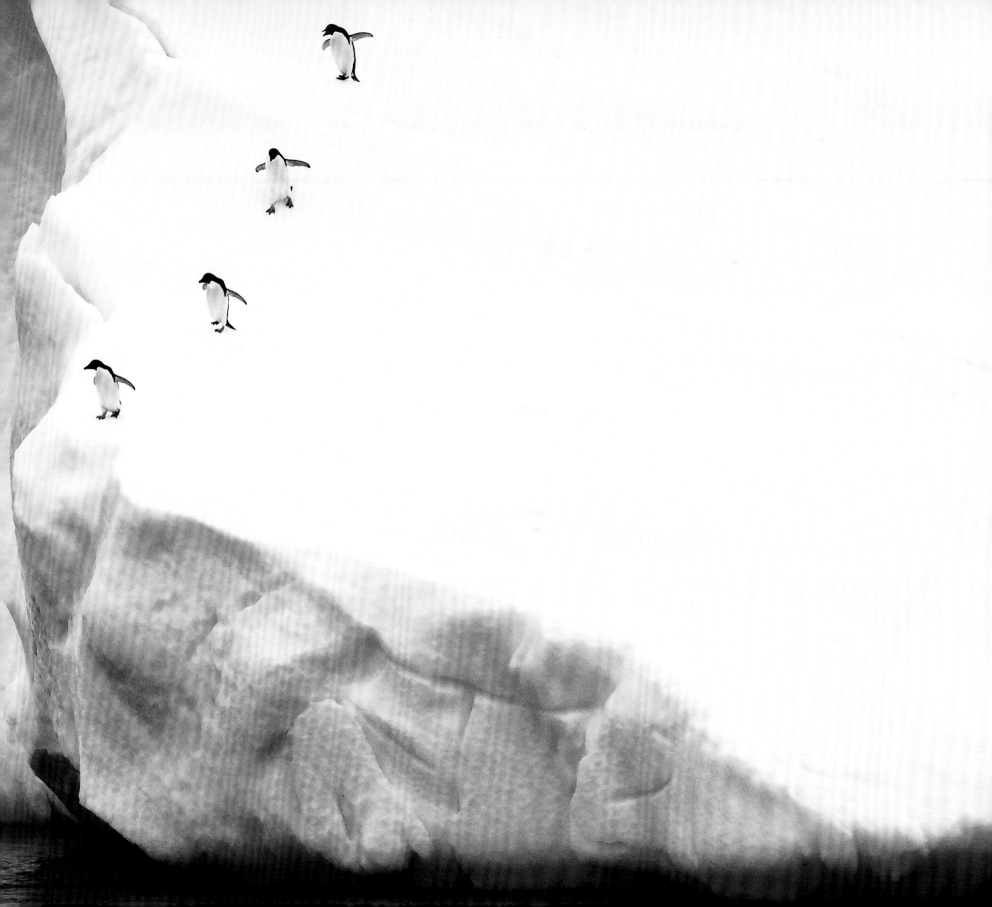

Nina Gallo is a writer, Antarctic historian and lecturer based in Australia. Also a certified Polar Tourism Guide and Zodiac driver, Nina has spent months working in Antarctic waters, and trained and worked as a remote first aider, canyon guide and rock climbing instructor. Nina completed her training as an expedition guide in the Blue Mountains, NSW and Hobart, Tasmania, where she also participated in a polar expedition medicine course run in conjunction with the University of Tasmania and the Australian Antarctic Division.
Nina contributes regularly to citizen science projects and volunteers with the NSW SES Bush Search and Rescue Unit. Her writing has been published in a number of publications.

CHAPTER 1

GREAT SOUTHERN LANDS

Australia and Antarctica share an ancient connection, which spans deep time, mythology and human endeavour. From prehistoric geological connections to early exploration and a flourishing Australian Antarctic Program today, the relationship between Australia and Antarctica is profound and enduring.

Previous: The Aurora Australis shimmers over East Bay, seen from Mawson Station. The aurora is the result of energetic charged particles colliding with gases in the atmostphere, such as oxygen and nitrogen, which emit light. This page: AAD staff and visitors arrive at Wilkins Aerodrome, about 70km south-east of Casey station and 3429km from Hobart. The foundation of the runway is glacial ice.

Geologist Douglas Mawson (centre) led the Australasian Antarctic Expedition, 1911–1914. The main base was established at Cape Denison on Commonwealth Bay, the departure point for his journey with Xavier Mertz and Belgrave Ninnis in late 1912. It was to be the most gruelling experience of Mawson's life, described in his riveting book, *The Home of the Blizzard*.

THE AUSTRALIAN ANTARCTIC TERRITORY

IT'S HARD TO say exactly when the connection between Australia and Antarctica began. Was it with Douglas Mawson, whose tales of polar exploration are taught in primary schools across Australia, and whose piercing gaze once graced the Australian $100 note? Or was it in 1804, when British explorer Matthew Flinders scrawled the name 'Australia' on a hand drawn map of our continent, unceremoniously snatching the title away from the as yet undiscovered Antarctica, the true Terra Australis Incognita? Perhaps it can be traced back billions of years, to when two embryonic continents collided, drifting together until they were torn asunder, becoming the Australia and Antarctica we know today.

Modern Australia's first direct contact with Antarctica came in the late 19th century when Louis Bernacchi, a young Tasmanian physicist, became the first Australian to set foot on the continent as part of the British Southern Cross expedition (1898–1900). It made sense for Australians to venture south. Hobart had long been a hub for mariners to rest, refuel and re-provision before sailing into the Southern Ocean.

In 1907 Ernest Shackleton was passing through Adelaide en route to Antarctica when he received a message from a 25-year-old geologist named Douglas Mawson. Mawson was fascinated by a vast accumulation of glacial sediments he had discovered in South Australia, and wanted to learn more about ice caps. He requested a berth for the return trip to Antarctica. Instead, Shackleton offered him a position as the expedition physicist. Mawson accepted.

A few years later Mawson returned to Antarctica as leader of the Australasian Antarctic Expedition $\longrightarrow$

Tasmanian Louis Bernacchi grew up on Maria Island and is commemorated in this statue on Hobart's waterfront. Bernacchi was physicist on Borchgrevink's 1898 expedition, the first expedition to endure winter on the Antarctic continent.

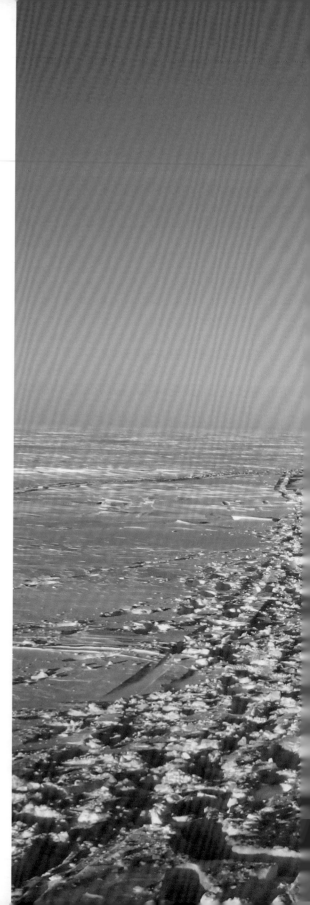

(AAE: 1911–13), the first Australian-led expedition to Antarctica. The expedition was plagued with difficulties from the start. Mawson established their Main Base in what he later described as "an accursed land," a bitterly exposed bay battered by ceaseless gales. Later, they suffered the tragic loss of two men, Xavier Mertz and Belgrave Ninnis, the two men who cared for the expedition's sled dogs.

Despite these challenges, the AAE was lauded as a success. It established the first radio connection between Antarctica and the outside world, and mapped vast swathes of previously uncharted land, providing a basis for future Australian Antarctic claims. Perhaps the expedition's greatest legacy was its comprehensive scientific data, which scientists still refer to today.

In 1929 Mawson returned to Antarctica as leader of the British Australian and New Zealand Antarctic Research Expedition (BANZARE), exploring more than 3000km of Antarctic coastline. The previous decade had seen a growing international interest in what had come to be known unofficially as the 'Australian Quadrant' of Antarctica, south of Australia. Mawson claimed sovereignty for Britain over the vast territory explored by BANZARE but it became the Australian Antarctic Territory (AAT) in early 1933 when Britain transferred its sovereignty to Australia. Covering nearly 5.95 million sq.km. – roughly 42 per cent of the Antarctic continent – it is by far the largest sector to be claimed by any nation.

Many expeditioners go to Antarctica to conduct field programs and spend a considerable part of their time living off-station. The polar pyramid tent can withstand winds of 100km/hour in full blizzard conditions.

Victoria Heinrich, meteorologist, releases a weather balloon with attached radiosonde at Mawson station. The balloon is launched every day of the year to record temperature, pressure, humidity, wind speed and direction.

THE AUSTRALIAN ANTARCTIC DIVISION

THE IDEA of a coordinated Australian scientific program in Antarctica was first suggested in 1886, but it wasn't until 1947 that the first research station was established on the subantarctic Heard Island. Australia's first Antarctic research station, Mawson, officially opened on 13th February, 1954.

As of May 2020, the Australian Antarctic Division (AAD) operates and maintains three permanent, year-round Antarctic research stations: Mawson (67°36'S), the longest continuously operating station south of the Antarctic Circle; Davis (68°35'S, est. 1957), the furthest south of Australia's stations; and Casey (66°17'S, est. 1969), known as the 'Daintree of Antarctica' due to its lush, extensive moss beds. The AAD also operates one subantarctic station on the World Heritage listed Macquarie Island.

AUSTRALIAN WOMEN IN ANTARCTICA

FOR MOST OF Australia's early Antarctic history, women were not welcome on Antarctic expeditions. In 1959, scientist and AAD secretary Susan Ingham approached Phillip Law, director of the AAD, to ask for a berth on a resupply trip to the station at Macquarie Island. Law had received applications from three other female scientists that year, and decided that since they would fill a cabin, they could go. The women were warned that on their "behaviour rested the future of [their] sex with regard to ANARE (Australian National Antarctic Research Expeditions)

voyages". Their pioneering research paved the way for female Antarctic scientists today. In the summer of 1960–61, artist and poet Nel Law, Phillip Law's wife, became the first Australian woman to stand on the Antarctic continent. It wasn't until 1975, International Women's Year, that Australian women were sent to work on the Antarctic continent as part of an AAD expedition. In 2017, women made up half of the overwintering team at Macquarie Island station. In 2020, across all stations, women comprised about a quarter of the Australian Antarctic population.

Elizabeth Chipman in 1976, one of the first three Australian women sent to work in Antarctica. The first female station leader on the continent, Diana Patterson, was appointed head of Mawson in 1989.

Following the 'blizz line' at Mawson station by a 300kW wind turbine. Most of the station's electricity needs are met by making use of the most powerful winds on the planet.

A row of Hägglunds after a blizzard. These Swedish over-snow vehicles can be driven on 50cm-thick sea-ice, carry four passengers and tow up to two tonnes on sleds.

STATION LIFE

SINCE 1948, THOUSANDS of ordinary Australians have lived and worked on our remote scientific stations in Antarctica. Their reasons are as varied as the individuals who apply. Some are looking for professional development or new skills. Others go to save money, experience a simpler life, or just for the adventure. Whatever their motivation, by the start of the 21st century, generations of scientists, plumbers, diesel mechanics, doctors, electricians and chefs had joined the ANARE Club for Australian Antarctic expeditioners. So, what is daily life like for those who travel south to Antarctica with the Australian Antarctic Program?

Utilitarian buildings litter a desolate blinding-white or dirty-grey landscape like gigantic Lego blocks. Some are converted shipping containers, insulated and painted in primary colours. As in all snowy environments, architecture that blends in is not a virtue here. Antarctic stations may not be beautiful, but they certainly get the job done.

Australian stations have comfortable living quarters, research laboratories, powerhouses, workshops and storehouses packed to the rafters with enough food, fuel and resources for expeditioners to be self-sufficient for months at a time. With up to 100 people on station during the summer, a single resupply can take between 10 and 14 days, all staff working shifts around the clock to unload countless shipping containers, machinery, vehicles and fuel, and ferry them from ship to shore.

Some people compare daily life on an Antarctic station to life on a mine site. It can be harsh and repetitive, far removed from the comforts and freedoms of home. Every expeditioner undertakes mandatory survival training and quickly learns to keep warm layers on hand and a weather eye out.

It's not unheard of for people to get lost only metres from a building in white-out conditions, and a network of ropes, known as 'blizz lines', is strung up across the station to guide people home in bad weather. When a blizzard closes in, visibility can get so low that no one is allowed to go outside at all.

Although Antarctica is bathed in 24-hour daylight during the summer, temperatures seldom rise above 5°C. When they do, the melt imposes its own restrictions, with the closure of soggy trails connecting the station to nearby huts and penguin colonies.

Despite its challenges, life in Antarctica has many charms. Everyone is encouraged to embrace station life, picking up a day of 'slushy' duty each month (a combination of dish pig, sous-chef, cleaner and DJ) or training as part of the search and rescue or firefighting crews. $\longrightarrow$

'I HAVE NEVER HEARD OR FELT OR SEEN
A WIND LIKE THIS. I WONDERED WHY
IT DID NOT CARRY AWAY THE EARTH.'

APSLEY CHERRY-GARRARD, 1911

The social calendar is filled with fancy dinners, dress-up parties and inter-station competitions, and most people take recreational trips (known as 'jollies') at every opportunity. For many, the experience of cross-country skiing to a bustling Adélie penguin colony, jumping in a Hägglund (an oversnow vehicle) for a trip to a remote hut or taking a Zodiac cruise among otherworldly icebergs is life-changing.

When the last resupply ship or flight leaves at the end of the summer, the 20 or so remaining expeditioners hunker down for the long, cold night. They will be literally cut off from the outside world by ice until next summer.

A typical wintering crew includes a station leader, medical officer, chef, diesel mechanics, communications and radio technicians, Bureau of Meteorology observers, plant operators and other tradespeople including plumbers, carpenters and electricians.

They settle into a routine, maintaining the station and assisting with year-round scientific observations.

Some expeditioners rate the midwinter swim as the highlight of the season, while others prefer the tradition of a lavish midwinter feast, or 'drive-in' movies projected on snow banks under the swirling colours of the aurora australis.

Winter can be a trying time in Antarctica. Temperatures can drop below −40°C, and Australian stations are plunged into months of twilight and weeks of complete darkness. There are stories of people breaking under the strain of living closely with others in confined quarters, away from family and friends. Despite its challenges, most overwinterers return home with lifelong friendships and fond memories of their winter in Antarctica.

Left: Davis station, home to about 100 expeditioners in summer. Davis sits on the coast of the ice-free Vestfold Hills.

Above: Expeditioner Paul Hanlon, wrapped up in multiple layers, lifts his goggles, briefly exposing bare skin to the cold, dry air.

This free-swimming robotic submarine is measuring the topography of the underside of a sea ice floe to learn more about its thickness and volume — crucial information for climate scientists.

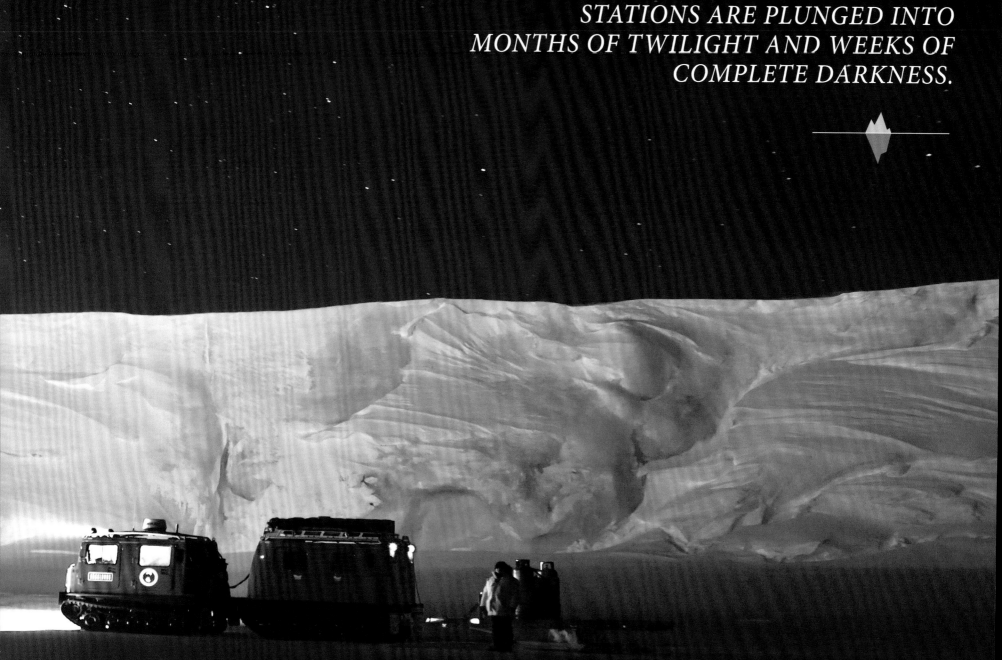

STATIONS ARE PLUNGED INTO
MONTHS OF TWILIGHT AND WEEKS OF
COMPLETE DARKNESS.

THE AAD INTO THE FUTURE

IN 2016, THE Australian Government unveiled their Australian Antarctic Strategy and 20 Year Action Plan. Between 2016 and 2036 the Australian Antarctic Program is set to expand as the Australian Government invests significantly in Australia's presence in Antarctica.

The iconic orange icebreaker, *Aurora Australis*, has been Australia's Antarctic flagship for 30 years. This much-loved research and resupply vessel is scheduled to be replaced by RSV *Nuyina* in early 2021. The new icebreaker is equipped with facilities and instruments for conducting scientific research in the Southern Ocean, and can break through sea ice 1.65m thick. Nuyina means 'southern lights' in palawa kani – the modern-day language of Tasmania's Indigenous population, which is based on spoken and written records of multiple Aboriginal languages that once existed on the island.

In May 2018 the Australian Government announced a plan for a new 2.7km paved runway near Davis station. If approved, the proposed Davis Aerodrome could provide year-round access to Antarctica by 2040, and will be the first paved runway in Antarctica. The existing Wilkins Aerodrome ice runway only operates in summer, with a 6-week hiatus – sometimes longer – through the warmest part of the season. The Davis Aerodrome would offer scientists unprecedented access to the continent, supporting research projects throughout the year.

The Australian Antarctic Program relies heavily on diesel fuel, and is eager to move towards more renewable energy sources. In 2003 the AAD introduced two wind turbines at Mawson, to provide up to 95 per cent of the station's energy requirements and save more than 600,000L of diesel fuel each year. Some field huts, weather stations and VHF repeaters are powered by solar energy. With frequent strong winds and 24 hours of daylight during the summer, there is enormous potential to harness renewable energy on Australian Antarctic stations in the future.

Servicing the automated seabird camera at Hawker Island, 7km southwest of Davis station. Southern giant petrels are studied remotely as they are nervous breeders and extra sensitive to human disturbance.

CHAPTER 2

EXTREME CONTINENT

Antarctica is a land of great extremes. Covered in a vast ice sheet almost twice the size of Australia, with winter temperatures that regularly fall below –60°C, Antarctica is the highest, coldest and windiest continent, and the largest desert on Earth.

Previous: Arcus shelf clouds near Casey station. This page: Overlooking the Olympus Range in Victoria Land. Fossils found in this Dry Valleys region provide evidence of changes in Antarctica's climate and environment. Lake deposits with perfectly preserved fossils of mosses, diatoms and minute crustacea, as well as beech leaf fossils found in the Friis Hills, reveal an abrupt cooling occured about 14 million years ago.

Right: Freeze-dried and mummified seal near Lake Vanda. Animals can lie undisturbed for thousands of years, preserved in cold dry air that prevents decay.

HOW ANTARCTICA WAS FORMED

WE LIVE ON a dynamic planet. Just beneath the Earth's surface is a tessellation of tectonic plates: terrestrial puzzle pieces gliding over a semi-molten mantle, carrying continents inexorably across the globe. Tectonic plates are the great creators and destroyers of our planet, giving rise to mountains, volcanoes, islands and causing earthquakes as they diverge and collide.

The origins of Antarctica can be traced back 3.8 billion years, to when the molten Earth was first beginning to cool. A layer of rocky crust encased the planet, swirling gases condensed into water to create the first oceans, and the earliest intimations of future continents took shape on a lifeless globe.

For millennia these primitive landmasses drifted amidst the primordial soup of ancient oceans, merging and tearing apart. About 500 million years ago a rocky mass collided with present-day India, Africa, South America and Madagascar to form an enormous supercontinent called Gondwana.

The supercontinent began to fragment 180 million years ago. The continents we know as South America and Africa separated first, with India and Madagascar breaking free about 40 million years later. The remains of Gondwana – the ancient core which would become Antarctica and Australia – continued its long, slow journey towards the south.

The Earth of the Cretaceous Period, 100 million years ago, would be unrecognisable to humans today. The polar regions were free of ice, sea levels were 200m higher and most of north-eastern Australia was covered by a vast inland sea. In this lush, tropical world, Gondwana drifted within the Antarctic Circle, where darkness prevailed for six months of the year and the brilliant southern lights of the aurora australis danced across winter skies. Gondwana was ice-free and teeming with life. Marine reptiles and invertebrates thrived in warm, coastal waters and the land was thickly covered with cool, temperate rainforest, where dinosaurs roamed.

Around 50 million years ago, a deep rift split Gondwana apart, creating the continents of Australia and Antarctica. The Australian plate began to move to the north, as it continues to do today. The geological connection between these continents, forged more than 3 billion years before, was over.

As Antarctica drifted further south, an immense ocean opened up, separating it from all other landmasses. Ferocious winds began to sweep across the sea, creating a fast, deep current encircling Antarctica like a moat. When the Earth's climate cooled around 34 million years ago, ice and snow accumulated on the mountaintops, creeping downhill as glaciers to form the Antarctic ice sheet. The Antarctica we know today was born.

The McMurdo Dry Valleys seen from a NASA satellite in 2002. Antarctica is the largest desert on Earth. The average yearly rainfall at the South Pole over the past 30 years was just 10mm. Scientists believe it took around 45 million years for the ice sheet to build to its current depth.

The Southern Ocean's Antarctic
Circumpolar Current connects
the Atlantic, Pacific and
Indian Oceans to form a global
network of ocean currents that
redistributes heat around the
Earth and so influences climate.

THE
SOUTHERN
OCEAN

ANTARCTICA AND THE Southern Ocean are inseparable. One begets the other, neither could exist in isolation. The Southern Ocean forms an invisible boundary around Antarctica, warding off the warm waters of the north and preserving the polar cold. To the south, the sea transmutes into ice, merging with the endless frozen streams emanating from the land, almost indistinguishable from Antarctica itself.

The Southern Ocean is a vast, circumpolar sea, an immense mixing bowl where all the world's oceans meet and intermingle. One of its most extraordinary features is the Antarctic Circumpolar Current (ACC), also known as the West Wind Drift. Covering a vast expanse of ocean up to 2000km wide and 4km deep, the ACC is the longest, strongest current on Earth. It flows unimpeded around the globe, charting a labyrinthine path through undersea mountain ranges and the constriction of the Drake Passage, south of South America, where the entire current is compressed into a channel only 800km wide. $\longrightarrow$

An Adélie penguin (*Pygoscelis adeliae*) looks for a gap in the sea ice. Adélies nest on land during the summer, and migrate during the winter to the edge of the sea ice, where they are able to feed at sea.

Sea ice averages about a metre thick but it can vary greatly – from 'fast ice' extending out from the coast, where the cover is often total, to looser pack-ice and polynyas – large areas of open water.

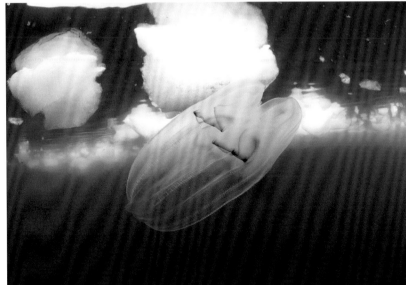

A ctenophore, or comb jelly, swims under the sea ice in search of planktonic prey.

The ACC flows west to east, transporting more water around the globe than any other current on Earth. This tremendous flow is initiated by two factors: the persistent westerly winds that rip across the southern latitudes, earning the names Roaring Forties, Furious Fifties and Screaming Sixties; and the sharp drop in ocean temperatures between the Equator and the South Pole.

Within the ACC there are several distinct regions or 'fronts' – zones of confluence where the temperature, salinity and density of the ocean suddenly shift. At the Antarctic Convergence, also known as the Polar Front, dense, cold water from the south sinks beneath comparatively light, warm water from the north.

This zone of turbulence, where nutrients are transported to the surface, is between 32km and 48km wide and forms a biological boundary between Antarctica and the outside world. As you continue south through the Antarctic Convergence zone, the sea surface temperature drops dramatically from 5.6°C to below 2°C, continuing to fall more gradually until it reaches –2°C.

When the sea temperature reaches –1.8°C it begins to freeze, at the surface and in typical salinity. Each winter, a vast expanse of sea ice radiates out from the Antarctic coast, covering 18 million sq.km. and almost doubling the size of the continent. Antarctic sea ice is a hydrological phenomenon of global significance, which has a major influence on the world's climate. Covering an area two or three times the size of Australia, its white surface reflects up to 90 per cent of the Sun's radiation (the ocean reflects about 5 per cent), helping to maintain the cold Antarctic climate. The textured underside of the sea ice is a complex three-dimensional landscape which provides critical habitat for microalgae and other organisms, and shelter for zooplankton.

Sea ice also has an important effect on deep ocean currents. As seawater freezes, some of the salt content is expelled into the ocean. This creates some of the saltiest, coldest, most dense water on Earth – Antarctic bottom water – which slides off the continental shelf, sinking into the ocean's abyssal depths. This high salinity and pressure at depth prevents ice forming, hence Antarctic bottom water can reach temperatures below 'freezing' and still be liquid. The momentum of this water as it slips along the seafloor acts as an oceanic engine, driving the global conveyor belt of currents that transport water around the world and regulate the Earth's climate.

A LAND OF ICE

Ross Ice Shelf, directly south of New Zealand, is named after Royal Navy captain James Clark Ross, who discovered it in January 1841 while in command of HMS *Erebus*.

TO APPROACH ANTARCTICA is to enter a world of brilliant white. Sea ice merges with seemingly infinite glaciers flowing from the land. Immense, frozen shelves of ice extend out from the coast, and marvellously sculpted icebergs sail on seas aglitter with gelid fragments sparkling in the sun.

The Antarctic Ice Sheet is the largest single mass of ice on Earth. This vast, frozen dome covers approximately 98 per cent of Antarctica, or 14 million sq.km. It is an awe-inspiring presence – at once timeless and transient, formidable and vulnerable.

The Antarctic Ice Sheet contains around 90 per cent of the world's ice and 70 per cent of its fresh water. Its thickness varies across the continent, reaching a maximum of 4776m in East Antarctica – more than half the height of Mt Everest. If it all melted, the ice sheet would contribute about 57m to global sea level rise. The bedrock beneath, currently compressed under the tremendous weight of this ice, would slowly rebound about 1000m.

The ice sheet is moving constantly from the higher middle part of the continent towards the coast, under the force of gravity and its own extraordinary weight. It travels about 10m each year at the South Pole, but enormous ice streams flowing over steeper terrain can travel up to 1,000m a year.

When the ice sheet reaches the coast it either calves into the ocean, becoming an iceberg, or continues to flow over the top of the sea, becoming a floating ice shelf. Ice shelves play an important role in helping to stabilise the Antarctic ice sheet, buttressing glaciers as they surge towards the sea. The largest ice shelf on Earth is the Ross Ice Shelf. Almost as large as France, it terminates in sheer ice cliffs up to 50m tall and hundreds of kilometres long. Away from its ocean edge, it is up to several hundred metres thick.

Each year Antarctica calves around 252 billion tonnes of ice into the ocean in the form of icebergs – roughly six times more than it did 40 years ago. The largest iceberg in recorded history calved from the Ross Ice Shelf in March 2000. Named Iceberg B-15, it was an extraordinary 295km long and 37km wide, larger than the Big Island of Hawai'i

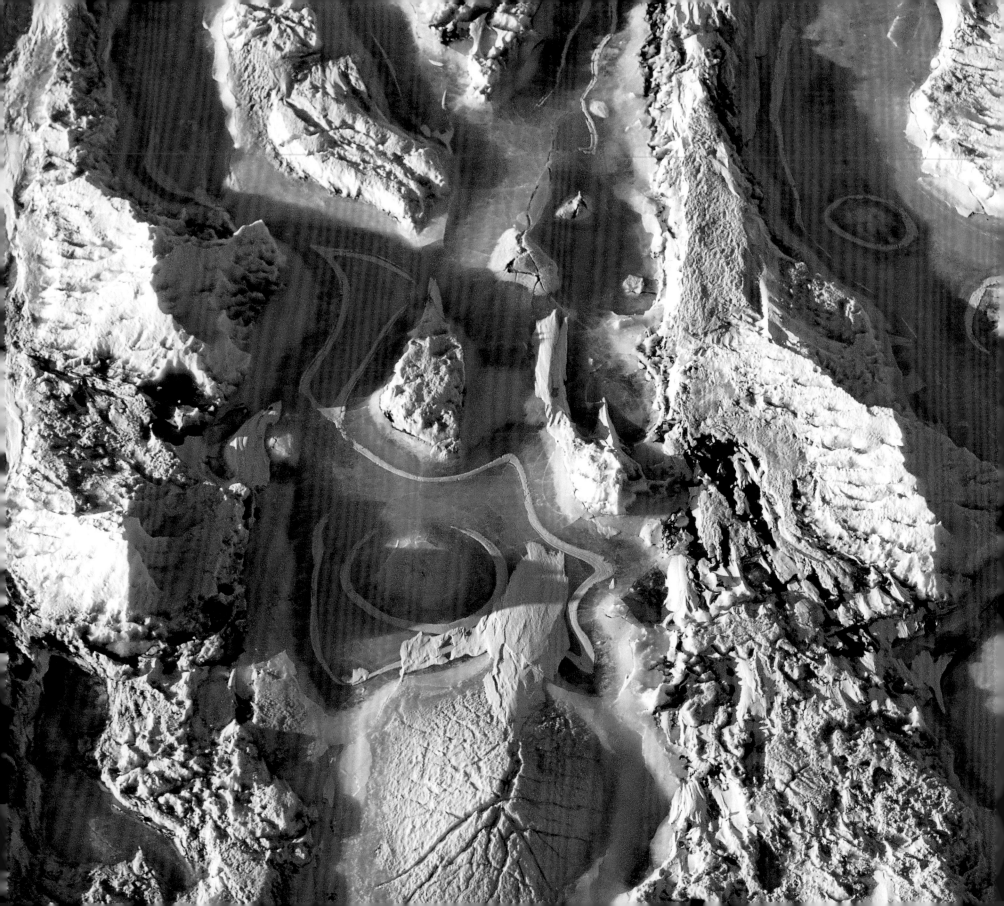

Opposite: Minna Bluff, at the southern end of the Trans-Antarctic Mountains, overlooking Ross Ice Shelf. Right: The Trans-Antarctic Mountains pierce the cloud layer, Mt Minto region.

BENEATH THE ICE

THE ANTARCTIC ICE Sheet covers an area almost twice the size of Australia. With the addition of other ice features such as tongues and rumples, this leaves a mere 0.4 per cent of Antarctica permanently ice-free. If we could peel back the ice and peer at the rocky topography underneath, what would we find? A continent? An island chain? This question has enthralled scientists for generations, but Antarctica's thick, icy mantle presents an almost insurmountable barrier. In recent years, however, new technologies have begun to reveal the complex and sometimes surprising landscapes hidden beneath the ice of the least understood continent on Earth.

The Trans-Antarctic Mountains are Antarctica's most striking geological feature: a towering string of summits, some rising more than 4,500m to pierce the surrounding ice fields. This chain of icebound peaks, known as nunataks, extends over 3,500km and forms a geographical border dividing Antarctica into two discrete sectors.

West (or Lesser) Antarctica is relatively young, generated by many of the same geological processes that created the Andes in South America. Between 500 and 35 million years ago a mosaic of tectonic plates converged and gave rise to the fault lines and folds, volcanoes and mountain ranges we see in West Antarctica today.

Much of the bedrock here lies below sea level, overlaid by the West Antarctic Ice Sheet. In the absence of this ice sheet, the ocean would flow hundreds of kilometres inland, inundating valleys up to 2400m deep, and much of what we recognise as West Antarctica would be submerged. Without its ice, West Antarctica would consist of several rocky archipelagoes dispersed across a vast sea, with the dramatic Trans-Antarctic Mountains rising to the east.

If we could cross the Trans-Antarctic Mountains and travel down their eastern slopes, we would look down over the awe-inspiring East Antarctica. $\longrightarrow$

Striking layers of dolerite between the sandstone strata of Finger Mountain in the McMurdo Dry Valleys. Antarctica's relatively small patches of exposed rock provide valued stamping grounds for scientists tracing the history of the continent.

*AROUND 50 MILLION YEARS AGO,
A DEEP RIFT SPLIT GONDWANA APART,
CREATING THE CONTINENTS OF
AUSTRALIA AND ANTARCTICA.*

The Matusevich Glacier flows toward the coast of East Antarctica, pushing through a channel between the Lazarev Mountains and the Wilson Hills. Fast ice anchored to the shore surrounds both the glacier tongue and the icebergs it has calved. Out to sea, the sea ice is thinner and moves with winds and currents.

Also called Greater Antarctica, this is a primordial place: the oldest rocks here are more than 3 billion years old, among the oldest still existing on Earth. Until recently, some scientists believed much of East Antarctica was a single continental mass sitting well above sea level. However, recent research revealed a more convoluted, fragmented story.

In 2019 scientists unveiled the most detailed Antarctic topographical map to date. High resolution images derived from years of satellite data, laser and ice-penetrating radar revealed the deeply textured surface of subglacial East Antarctica. Several glacier beds were found to be hundreds of metres deeper than previously thought. Further, scientists realised that the massive Denman Glacier, which plunges

to 3500m below sea level, is the deepest terrestrial canyon on Earth.

These new data also revealed that East Antarctica is not a single continental mass, but instead what some have termed a "graveyard of continental remnants…the wreckage of an ancient supercontinent's spectacular destruction". These rocky fragments, called cratons, share some geological traits with Australian and Indian bedrock. One section shares many similarities with the Mawson Craton, part of which extends into Southern Australia – a remnant of a shared Gondwanan past. Exactly where and when these cratons originated is part of an ongoing puzzle, a subject of future research for Antarctic geologists.

CHAPTER 3

LIFE ON ICE

Antarctica is home to a community of
remarkable wildlife uniquely adapted to thrive in
an environment many would consider inhospitable.
From microscopic invertebrates on land to
charismatic penguins and enormous blue whales
at sea, the Antarctic ecosystem tells a tale of
extraordinary adaptability and resilience in some of
the harshest conditions on Earth.

Previous: This Adélie penguin (*Pygoscelis adeliae*) has a pinkish underside to the wings, indicating active bloodflow from recent swimming activity.
This page: Weddell seals (*Leptonychotes weddellii*) produce a larger range of sounds than any other seal. Weddell 'songs', repetitive sequences of those sounds, have been recorded.

SEALS

THERE ARE SIX seal species living in Antarctic waters: elephant, Antarctic fur, leopard, crabeater, Weddell and Ross seals.

Weddell seals are known for their lovely, mottled coats and beautiful singing voices. Crabeater seals don't eat crabs, but krill. They are predominantly filter feeders, and have special lobed (or lobodontine) teeth, which act as a sieve, allowing them to scoop up a mouthful of krill and seawater, then push the water out.

Ross seals (*Ommatophoca rossii*) are seldom seen, and have enormous eyes up to 7cm in diameter, while Antarctic fur seals are quite gregarious, often hauled out in groups on rocky beaches in the northern reaches of Antarctica. Leopard seals are very solitary apex predators. While their diet consists mainly of krill, they supplement this with penguins, and even small seal pups. Southern elephant seals are the largest seals on Earth, growing to almost 7m long and weighing over 4000kg.

All Antarctic seals are covered in a thick layer of blubber, or fat, which helps fend off the cold. Fur seals also have highly specialised fur, with two layers to insulate and protect.

Antarctic seals spend most of their lives at sea or resting on ice floes, although some haul out to sunbathe on rocky beaches. They feed primarily on krill, fish and small crustaceans, and have large eyes and whiskers (vibrissae) to help them home in on their prey in dark polar waters.

Seals are excellent swimmers and divers. Recent research using electronic transmitters revealed that some elephant seals migrate an extraordinary 5482km annually, diving to 2389m and staying underwater for 94 minutes.

Elephant and fur seals generally breed on subantarctic islands, including South Georgia and Macquarie Island. The other species breed among the Antarctic pack ice. Relatively little is known about their breeding cycles as they don't appear to congregate in regular colonies, making it difficult to observe them consistently. Only the Weddell seal returns to the same breeding site to rest, moult and give birth to pups on the ice. Females older than 6 years generally give birth to one pup in October. The only known observation of Weddell seal mating, which took place in 1971, reported that it took place underwater.

Most Antarctic seals migrate to alternative feeding grounds during the winter. The Weddell seals are the one exception. The southernmost seals on Earth, they live on and around sea ice stuck fast to the shore, rarely venturing far from their birthplace. Crabeater seals tend to migrate with the sea ice edge as it expands and contracts, while elephant seals, Ross seals, some leopard seals and Antarctic fur seals drift between Antarctica and subantarctic islands throughout the year. While it's rare for Antarctic seals to venture north of the subantarctic, vagrants of all species have been spotted as far north as Australia.

Antarctic fur seals (*Arctocephalus gazella*) do not have the same thick layer of fat as true Antarctic seals, and are distinguished from them by visible ear flaps. They breed mainly on the islands south of the Antarctic Convergence, like these ones in South Georgia.

Top left: Flipper of Southern elephant seal (*Mirounga leonina*). Top right: Leopard seal (*Hydrurga leptonyx*).
Bottom: A harem of Southern elephant seals on South Georgia. Following: Crabeater seals (*Lobodon carcinophaga*)
often have long scars down the sides of their bodies, most likely inflicted by their main predators – leopard seals.

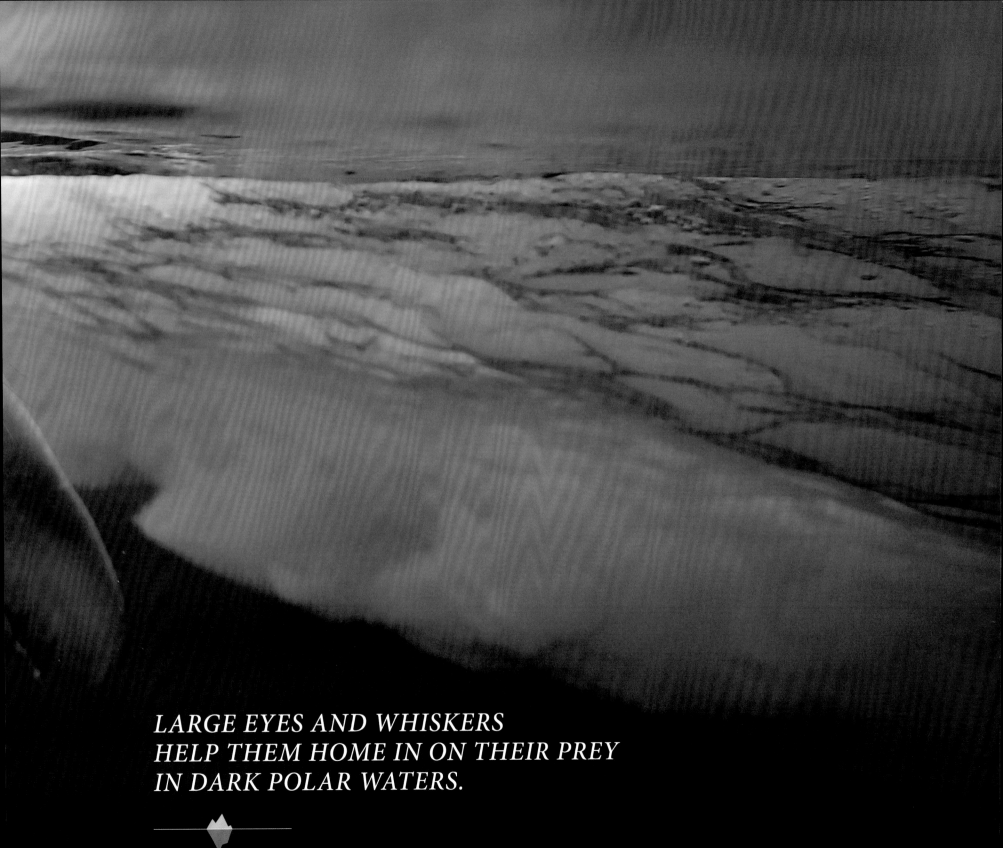

*LARGE EYES AND WHISKERS
HELP THEM HOME IN ON THEIR PREY
IN DARK POLAR WATERS.*

PENGUINS

OF APPROXIMATELY 18 penguin species on Earth, there are six that live and breed on the Antarctica continent: Adélie, emperor, chinstrap, gentoo, rockhopper and macaroni.

Early Antarctic explorers often wrote about penguins with fondness as "the most comical caricature imaginable of a stout, elderly gentleman, in a dress-suit…at once a little conceited and very dignified". While penguins are known and loved for their humanesque qualities on land, they are most at home in the ocean. Some explorers who encountered penguins in the water mistook them for fish.

These flightless seabirds are sleek, streamlined swimmers that live much of their lives at sea, where they feed primarily on fish, krill and other crustaceans. Their distinctive plumage, white on the front and black on the back, serves as camouflage to protect them from oceanic predators such as killer whales and leopard seals.

Penguins stay warm with a thick layer of blubber, or fat, and dense, overlapping layers of down and feathers. They keep these waterproof by coating them with preening oil, which comes from a gland above their tail. Each year penguins undergo a catastrophic moult, when they shed the previous year's plumage, replacing all their feathers for the upcoming winter.

Emperor penguins, the largest of the penguin species, stand at around 1.2m tall and weigh up to 40kg. Emperors are the only penguins that breed in the Antarctic winter. The female lays an egg at the end of summer, leaving her male partner to keep it warm for two months while she goes off to forage for krill, fish and squid. The males fast and huddle to retain warmth in the dark and cold until the female returns to care for their newly born chick.

Adélie penguins join emperors as the southernmost penguins on Earth. Also found only in Antarctica, they grow to around 70cm in height and are known for their lustrous plumage and the white rings around their eyes. They typically breed in large colonies on rocky outcrops near the coast, building nests out of pebbles, which are also gifted to attract possible mates. In a busy rookery, pebbles are a hot commodity. Adélie penguins, along with their gentoo and chinstrap friends, are known for pilfering pebbles from neighbouring penguins to build their nests and impress potential partners.

Penguin populations are shifting in the Antarctic Peninsula, one of the most rapidly warming areas on Earth. Adélie and chinstrap colonies appear to be declining, with gentoo penguin colonies moving in to fill the void. Adélie colonies continue to flourish in other locations across Antarctica and the chinstrap penguins in the subantarctic, and are currently not classified as threatened.

These emperor penguin (*Aptenodytes forsteri*) chicks are about 2–3 months old. Always hungry, at this age both parents need to go out and gather food for their youngster. It takes about five months to rear an emperor penguin chick.

Snow petrels (*Pagodroma nivea*) like to nest in deep rock crevices with overhanging protection.

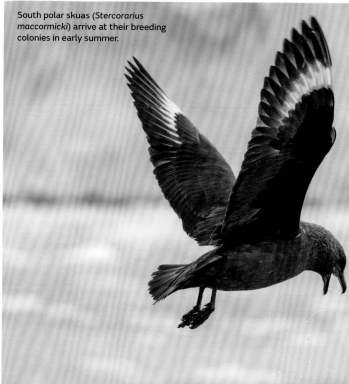

South polar skuas (*Stercorarius maccormicki*) arrive at their breeding colonies in early summer.

Southern giant petrel (*Macronectes giganteus*) chicks can fly at about 100–130 days old.

Arctic terns (*Sterna paradisaea*) are long-lived birds, commonly living from 15–30 years of age.

SEABIRDS

THE COLD, nutrient-rich waters of the southern latitudes are a seabird's paradise. Each spring more than 100 million birds flock to the rocky coastlines of Antarctica and nearby subantarctic islands, including albatrosses, petrels, skuas, cormorants, gulls, terns and, of course, penguins.

The pure-white snow petrel lives almost exclusively among the ice, and is one of only three bird species recorded at the South Pole. With a wingspan of approximately 80cm, it is among the smaller of the Antarctic seabirds, but its small stature belies its incredible grit. The snow petrel breeds further south than any other bird, except perhaps the south polar (McCormick's) skua. These graceful petrels find shelter amongst exposed rocks to build their nests of pebbles, some as far as 440km from the coast, in areas where temperatures fall below −40°C.

The south polar skua is another of the very few species that breeds in Antarctica, and is an occasional winter visitor to Australia. South polar skuas prey on penguin chicks and eggs, and are often seen loitering around colonies during breeding time.

The southern giant petrel is the largest petrel and has a wide range, from Australia to subantarctic islands and Antarctica. Sometimes called 'stinkpots' because of a smelly oil they can regurgitate and project up to a metre if disturbed, their diet consists primarily of krill, squid, fish and other small seabirds, including penguin chicks.

The Wilson's storm petrel (*Oceanites oceanicus*) is one of the smallest and most numerous seabirds in the world. With a wingspan of approximately 40cm, it feeds by skipping lightly across the sea surface, wings outstretched, scooping up small marine organisms with its beak.

One of the most remarkable visitors to the region is the Arctic tern. Each year it makes a round trip of about 90,000km from nesting sites in the Arctic to feeding grounds in Antarctica – the longest migration on Earth.

WHALES

EACH SUMMER THOUSANDS of baleen whales make the long migration from their breeding and calving grounds in the lower latitudes to Antarctica, to feast on the abundant swarms of krill for which the Southern Ocean is known.

Perhaps the most celebrated of Antarctic whales, the baleen species include the southern right, Antarctic minke, humpback, sei, fin, and blue whales. These whales are among the largest species on Earth. The blue whale, which can grow to 30m long and weigh nearly 200t, is the largest animal known to have ever lived on Earth. Fin whales come in a close second, growing up to 26m long.

Many Australians are likely familiar with the smaller humpback whales (12–16m), which migrate annually up the eastern and western coasts of Australia to their breeding grounds in the north. They travel a distance of about 10,000km, treating thousands of patient whale-watchers to shows of fin-slapping, spy-hopping and breaching along the way.

Despite their size, baleen whales feed predominantly on tiny Antarctic krill. Baleen are strong plates made of keratin, like our fingernails, which they have instead of teeth. When these whales approach a swarm of krill they scoop up huge quantities of krill and water, using their baleen as a filter to expel seawater as they swallow their prey. A blue whale can eat up to 40 million krill per day, more than 3500kg of the tiny crustaceans.

There are four species of toothed whale in Antarctica – sperm, southern bottlenose, the orca (killer) and Arnoux's beaked whale. Rather than filter feeding like the baleen whales, they catch single prey, ranging from fish and squid to penguins, seals and, in some cases, other whales.

Only one species of dolphin, the hourglass, is commonly found in Antarctic waters, although the southern right whale dolphin visits occasionally.

Antarctic whale populations are still recovering from decades of hunting throughout the 19th and 20th centuries. While some populations, such as humpback whales, appear to be strongly rebounding, others, like blue whales, are still much diminished. The International Whaling Commission estimates that more than 300,000 blue whales were killed in the Southern Ocean.

Antarctic Minke whales (*Balaenoptera bonaerensis*) are a baleen species that breed in tropical and subtropical waters, and migrate to Antartica to feed on krill. They grow to 10m and live for 30–50 years.

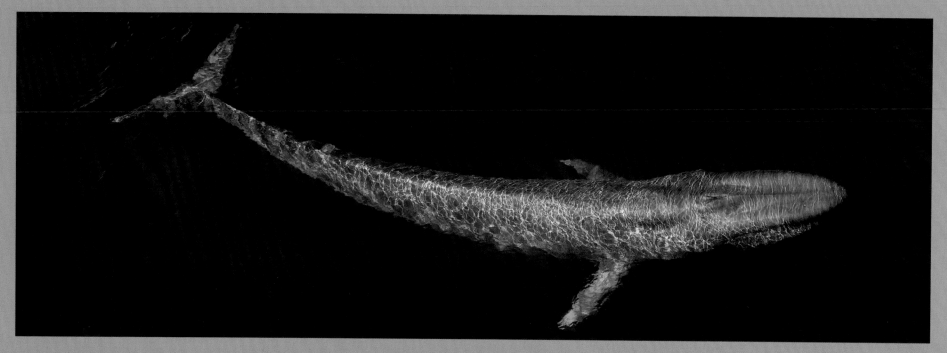

Top: Blue whale (*Balaenoptera musculus*), the loudest animal on Earth. Their calls reach 188 decibels. Left: The 1.8m-long hourglass dolphin (*Lagenorhynchus cruciger*) reaches speeds of 22km/hr and typically makes a lot of spray when surfacing to breathe.
Right: Orca whale (*Orcinus orca*) hunting a Gentoo penguin (*Pygoscelis papua*). There are thought to be about 70,000 orcas in Antarctic waters.

Southern Ocean humpback whales (*Megaptera novaeangliae*) fatten up on krill in Antarctica before undertaking one of the longest annual migrations of any mammal, travelling up to 8000km north to their tropical breeding grounds.

KRILL

ANTARCTICA has relatively few different species compared to the warm, tropical regions towards the Equator. But what it may lack in diversity, Antarctica makes up for in sheer abundance. There are an estimated 500 trillion Antarctic krill (*Euphausia superba*) in the Southern Ocean, weighing a total of 379 million tonnes – the largest biomass on Earth for a single species.

Krill are small crustaceans with large, black eyes and many small legs called thoracopods. They can grow up to 6cm long, and with a reddish tinge to their shell they resemble shrimp.

Krill may be small, but they have a huge impact. They are a keystone species in the Antarctic ecosystem, providing essential nourishment for millions of Antarctic inhabitants from fish and seabirds to penguins, seals and whales.

Two flowering plants – Antarctic hair grass (edge, *Deschampsia antarctica*) and Antarctic pearlwort (centre, *Colobanthus quitensis*) – are found on the northern and western parts of the Antarctic Peninsula.

Casey station is known as the 'Daintree of Antarctica' for its extensive moss, lichen, liverwort and algae communities. Casey's verdant moss beds are showing signs of stress due to the drying of East Antarctica, caused by both climate change and ozone depletion.

ANTARCTIC EXTREMOPHILES

ANTARCTICA PLAYS host to a profusion of extraordinary extremophiles, with impressive strategies for tolerating some of the harshest conditions on earth.

Antarctica has only two flowering plants, Antarctic hair grass and Antarctic pearlwort, but lush communities of moss and lichen flourish in coastal areas. Antarctic mosses grow at a glacial pace – typically just 1mm a year. These miniature forests are home to hardy, microscopic invertebrates like springtails and nematodes, which live in Antarctica all year. Minuscule tardigrades, also known as moss piglets, are almost indestructible inhabitants. They can survive temperatures as low as –200°C by dropping into a state of suspended animation, called cryptobiosis. They can maintain this state for 30 years, perhaps more, before coming back to life.

There are more than 100 species of fish in Antarctic waters close to the continent, 90 per cent of which belong to the sub-order Notothenioidea. These fish, including the Antarctic toothfish and crocodile ice fish, contain anti-freeze proteins that prevent their blood and bodily fluids from forming ice crystals in sub-zero temperatures.

CHAPTER 4

HUMAN ENDEAVOUR

"A plunge into the writing storm-whirl
stamps upon the senses an indelible and awful
impression seldom equalled in the whole gamut
of natural experience…the merciless blast – an
incubus of vengeance – stabs, buffets and freezes;
the stinging drift blinds and chokes."
Sir Douglas Mawson, 1914

IN THE BEGINNING

ANTARCTICA HAS EXISTED on the periphery of human consciousness for more than 2000 years.

When Ancient Greek philosophers began to realise that the Earth was round, not flat, they imagined a vast, southern land – a 'counterweight continent' to balance the land in the north. European cartographers sketched its hypothetical shores on maps and named it Antarktos (opposite Arktos, as the Arctic was known), or Terra Australis Incognita (Unknown Southern Land), but it would be centuries before anyone sailed south of the Equator to test their theories.

Some say that the Polynesians were the first to sail into Antarctic waters. A Maori legend tells of Ui-te-Rangiora, a great navigator from the Cook Islands who sailed with a fleet of waka (open canoes) in 650CE, discovering Te tai-uka-a-pia (frozen sea like arrowroot). They must have sensed the icy exhalations of their southerly neighbour. Did the palawa and pakana peoples of Tasmania also tell tales of a distant, frozen land, perhaps the birthplace of liyanana (ice) and parathiyana (snow)?

When Europeans finally crossed the Equator in the 1400s, they discovered that Terra Australis Incognita was not where the maps suggested. Magellan, Drake, Tasman – each returned with stories of sea where the great southern continent should be. People began to wonder whether it existed at all.

In 1772, British Captain James Cook and his crew forged a perilous path through a jostling labyrinth of "ice islands", sailing further south than anyone before. They crossed the Antarctic Circle three times and sailed more than twice the distance around the Equator, but they did not find Terra Australis Incognita. When they returned to Britain, Cook declared that if it existed, it would be so desolate that "the world [would] derive no benefit from it."

Previous: Mawson's Hut at Cape Denison, winter base for the Australian Antarctic Expedition led by Douglas Mawson. This historic and fragile wooden hut and others nearby are maintained by the Mawson Huts Foundation. Opposite: On Johannes Janson's map of about 1640, cautious conjectures of coastlines in green demonstrate the continuing belief in the existence of the Antarctic continent.

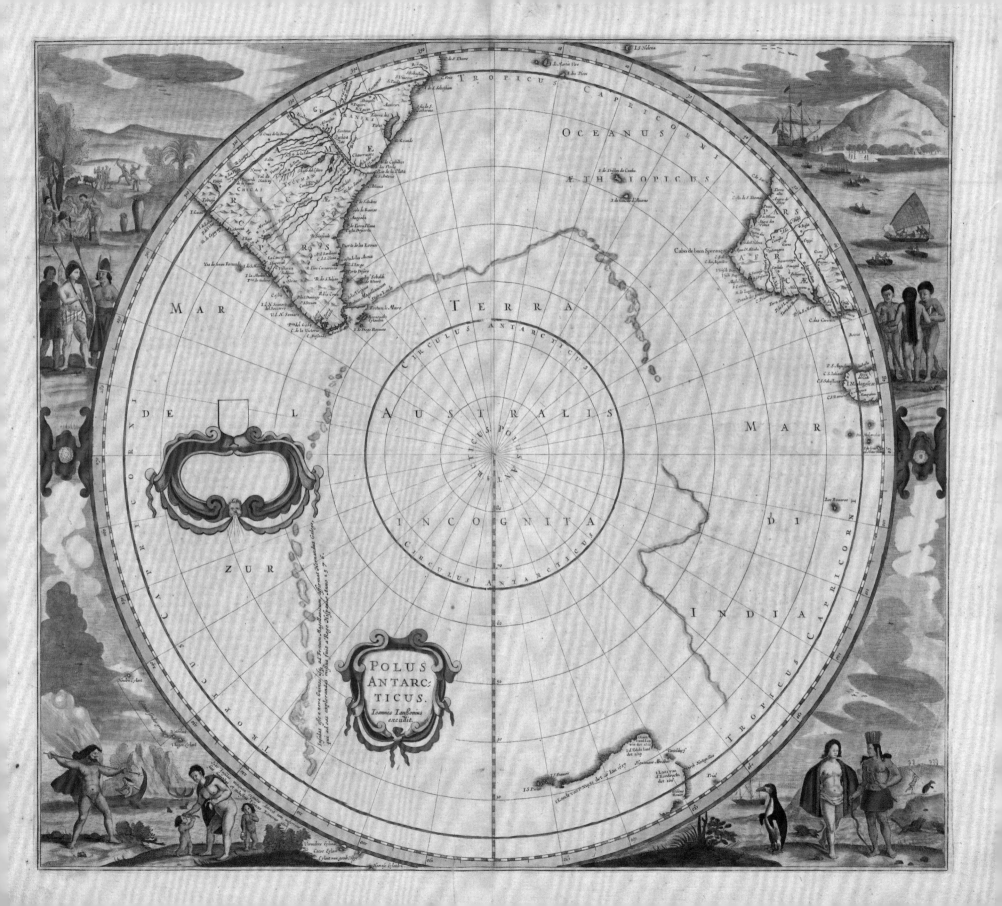

POLUS
ANTARC-
TICUS.
Ioannes Ianssonius
excudit.

This illustation from 1846 depicts Dumont D'Urville's *Astrolabe* and *Zélée* discovering Terre Adélie in January 1840. Terre Adélie, and the penguin species discovered by the same expedition, were named in honour of Dumont's wife Adèle.

AT FIRST SIGHT

HISTORIANS LOVE TO debate exactly who first sighted Antarctica but we will probably never know. What we do know is that in January 1820, after millennia spent imagining, hypothesising and exploring, human eyes caught their first glimpse of Antarctica. It may have been a member of the Russian Antarctic Expedition, led by Fabian Gottlieb von Bellingshausen, which spotted land while retracing Cook's circumnavigation. Perhaps it was a sailor aboard an exploratory British voyage to the Antarctic Peninsula, led by Edward Bransfield, an Irish-born officer of the Royal Navy. Or maybe it was a sealer, seeking out new colonies around the nearby South Shetland Islands.

The next decades belonged to the sealers. Cook's charts of the subantarctic islands were like treasure maps to sealers. Sailing from Britain, North America and South America, the exact details of their furtive expeditions remain shrouded in mystery. Few records were kept and limited traces remain. But by the turn of the 20th century, Antarctic fur seals were almost extinct and sealing was replaced by whaling, which continued in some form until as recently as 2019.

While most expeditions in the 19th century were commercial, the late 1830s saw a flurry of more exploratory national naval expeditions. French captain Dumont d'Urville discovered and claimed Terre Adélie, a slender wedge of Antarctica that divides the present Australian Antarctic claim in two. American Charles Wilkes and his crew sailed more than 1300km of uncharted eastern Antarctic coastline, and British captain James Clark Ross discovered the Ross Sea and Ross Ice shelf, also naming Mt Erebus, the southernmost active volcano on Earth.

At the dawn of the 20th century, we still knew very little about Antarctica. While a few expeditions had mapped sections of the coastline and set foot on the continent, the interior remained as mysterious as the Moon. No one had spent a winter on the mainland or taken an extended journey on the ice.

"IN THIS ICE FIELD WE COUNTED
NINETY SEVEN ICE HILLS OR MOUNTAINS,
MANY OF THEM VASTLY LARGE"

CAPTAIN JAMES COOK, 1774

THE HEROIC
AGE OF EXPLORATION

THE EXPEDITIONS that took place between 1897 and 1922 were defined by bold experimentation, great uncertainty and incredible hardship and sacrifice. Nineteen men lost their lives, and by the end many of Antarctica's great geographical landmarks had been explored.

In 1898 Anglo-Norwegian explorer Carsten Borchgrevink led a team of 10 men (including Australian physicist Louis Bernacchi) on a visionary but largely forgotten expedition. They became the first to overwinter on the continent, set foot on the polar plateau and journey towards the South Pole (albeit only a few kilometres, to 78°50′S), but their achievements were quickly eclipsed by the extraordinary journeys that followed.

Hundreds of men participated in expeditions to Antarctica through the early 20th century, but only a few have been immortalised. Shackleton, Scott, Amundsen, Mawson – these men, and their tireless pursuit of scientific understanding and the

geographic South Pole, have become synonymous with endurance, resilience and the 'heroic age' itself.

Robert Falcon Scott was appointed leader of the British Discovery expedition to the South Pole between 1901 and 1904. His third officer, Ernest Shackleton, was one of three men Scott selected to join him on his first attempt to reach the Pole. Antarctica was a formidable opponent. Trudging across the elevated glaciers of the polar plateau, the men were afflicted by furious winds, blinding white-outs and tortured towers of sastrugi (wind-blown ice features that can be metres high). Hampered by a lack of knowledge or experience of polar travel, their equipment was no match for the harsh environment, and they turned back after attaining a new southern record of 82°17′S.

Shackleton returned to Antarctica as leader of the Nimrod expedition in 1907. With him was a young Australian geologist named Douglas Mawson. While Shackleton prepared for his second push for the Pole,

Mawson joined small, multinational groups that became the first to climb Mt Erebus and reach the South Magnetic Pole.

Shackleton, demonstrating extraordinary restraint, turned his team around less than 160km from the Pole after an arduous journey of 73 days. Exhausted, malnourished and short on food, he recognised that the Pole, while deeply alluring, could cost them their lives. He returned to safety, telling his wife "I thought…you would rather have a live ass than a dead lion".

Norwegian explorer Roald Amundsen never intended to participate in a race for the South Pole. After enduring the first expedition to overwinter in Antarctica in 1898, he spent almost three and a half years making the first traverse of the Northwest Passage (1903–06). He and his crew overwintered with Inuit communities, learning valuable lessons about polar survival. In 1910 he was planning an expedition to the North Pole when he received $\longrightarrow$

Getting ice in a blizzard for domestic purposes, from the glacier adjacent to the hut, AAE 1911-14.

news that two Americans claimed to have reached it first. Eager to attempt a true geographical 'first', he slipped quietly south.

At the same time, Mawson was planning his next Antarctic voyage. He had heard Scott was also planning a second voyage and requested a berth on Scott's ship to launch an independent scientific expedition. Scott offered Mawson a place on his South Pole team, which Mawson fatefully declined.

When Scott set sail as leader of the Terra Nova expedition in June 1910, he was confident he had taken every possible measure to ensure its success. They had ponies, sled dogs, motorised sledges and ample provisions. But when Scott arrived in Melbourne there was a surprise waiting for him – a telegram from Amundsen reading "Beg inform you Fram proceeding Antarctic. Amundsen".

Seventeen years earlier, Norwegian polar explorer Fritjof Nansen had sailed the *Fram* on his attempt to reach the geographic North Pole. Now Amundsen was sailing the ship south. Scott knew exactly what this meant.

On 19 October 1911 Amundsen and his team of five men, four sledges and 52 sled dogs left their camp on the Ross Ice Shelf, about 10 days before Scott's did. On 14 December 1911 they reached the Pole. "Everything went like a dance," Amundsen reported. "We had a celebration dinner: a small piece of seal meat each." On 7 March 1912 Amundsen sailed the *Fram* into Hobart's harbour and sent telegrams to announce his victory to the world.

By contrast, Scott's expedition was plagued with troubles. They established their camp on Ross Island, 96km further from the Pole than Amundsen. The motor sledges failed within days and it wasn't long before the ponies floundered too. When they reached the South Pole on 17 January 1912, to find a Norwegian tent standing proudly on the icy, white plain, they felt thoroughly defeated.

"The POLE!" Scott wrote. "Yes, but under very different circumstances from those expected. Great God! This is an awful place and terrible enough for us to have laboured to it without the reward of priority."

They were 33 days too late, and their return journey would go down in history as one of the great tragedies of polar exploration. Disconsolate and struggling against unseasonably harsh weather,

all five had perished by the end of March 1912. Scott and three others were found, eight months later, entombed in their tent, devastatingly close to their final supply depot only 18km away.

After declining Scott's invitation to join the expedition to the South Pole, Mawson launched the Australasian Antarctic Expedition to explore Adélie Land, King George V Land and Queen Mary Land. Mawson and 31 expeditioners, most of them Australian, had departed Hobart in July 1911, three months before Scott's fateful journey to the Pole. In November 1912, Mawson, Ninnis (England) and Mertz (Switzerland) began an exploratory journey that would end in tragedy. In December Ninnis fell into a crevasse, and in January Mertz succumbed to severe illness. Mawson found himself alone in a heavily crevassed wilderness, more than 150km from help. His incredible 30-day solo journey to the coast would go down in polar history as one of the most remarkable stories of survival. That is, until Shackleton's expedition in the *Endurance*.

After Scott's tragic expedition to the South Pole, British support for Antarctic expeditions was at a low ebb. Winston Churchill, First Lord ⟶

Above: Roald Amundsen wrote in his diary before attempting the pole, 'The English have loudly and openly told the world that skis and dogs are unusable in these regions and that fur clothes are rubbish. We shall see — we shall see'. Top right: Essential supplies from 1911 left in situ, Mawson's Hut. Bottom right: Ninnis and Mawson (right) on the deck of the SY *Aurora* on the 1911 voyage down to Antarctica.

Left: Dr Xavier Mertz displays formation of an ice-mask. Mertz was to die on a gruelling traverse with Mawson in early 1913. Below: Mertz and his colleague Belgrave Ninnis, who died on the same journey, were remembered with a timber memorial cross erected near Mawson's Hut in November 1913. This replica cross overlooks Commonwealth Bay.

of the Admiralty, asked Shackleton, "What's the use of another expedition? Enough life and money has been spent on this sterile quest".

However, Shackleton believed that a successful expedition could help restore British national pride, and proposed the Imperial Trans-Antarctic Expedition, with the goal of completing a full traverse of Antarctica.

Disaster struck before the *Endurance* made landfall in Antarctica. Shackleton and his 24 men spent eight months trapped in sea ice before their ship was crushed by the ice. After five and a half wretched months camping on the treacherous, frozen surface of the sea, they escaped to a desolate island in three small lifeboats. Recognising that rescue from here would be impossible, Shackleton and five men completed a perilous, 17-day crossing to South Georgia, 1300km away. Their journey, in a wooden boat only 7m long, is one of the most remarkable small-boat voyages ever made. Within three months of their arrival in South Georgia, every man was rescued, returning to a world in the grip of World War I.

After the war had passed, Shackleton returned to Antarctica one last time. En route, anchored off the shores of South Georgia, he suffered a heart attack and was buried on the shores of a Norwegian whaling station in 1922. Shackleton's death marked the end of the Heroic Age of Antarctic Exploration.

Ocean Camp, 15 December 1915. Shackleton and his men were forced to set up camp on the ice after the *Endurance* sank on November 21. Many long months of misery lay ahead. They were not rescued until 30 August 1916.

Sir Edmund Hillary (left) and Vivian 'Bunny' Fuchs stand on the tracks of a Tucker Sno-cat B on the sea-ice outside Scott Base, after the completion of the Trans-Antarctic Expedition in 1958.

Vivien Fuchs at Plateau Depot, during the Commonwealth Trans-Antarctic Expedition, 1956-8.

Lowering a Lockheed Vega Float Plane from the ship at Port Lockroy during the Wilkins Hearst Antarctic Expedition, 1928-9.

MECHANICAL AGE OF EXPLORATION

THE EXPEDITIONS that followed the Heroic Age were of an entirely different nature, informed by 20 years of hard-won experience, cumulative knowledge and technological advancements.

On 16 November 1928 Hubert Wilkins, an Australian explorer, pilot, soldier and photographer, and American pilot Carl Ben Eielson made the first aeroplane flight over Antarctica. In December they made a 10-hour journey along the Antarctic Peninsula in a Lockheed Vega monoplane, mapping 970km of uncharted territory. This was followed by the first flight over the South Pole by Richard E. Byrd in 1929. During the 1930s and 1940s, Antarctic exploration slowed dramatically as ships and men were sent north to assist with the war effort.

Shackleton's vision for an Antarctic traverse was finally realised by the Commonwealth Trans-Antarctic Expedition (1956-58). Vivian Fuchs (England) led a 12-man team on the traverse, equipped with six vehicles, including Sno-Cats and modified tractors, and supported by a group led by Edmund Hilary (New Zealand). They completed the historic 99-day, 3473km crossing on 2 March 1958, a feat not repeated for another 23 years.

Fluctuations in sea-ice are being studied for their impact on the Antarctic ecosystem and global climate.

THE RISE OF SCIENCE

WHILE MOST early Antarctic expeditions incorporated scientific programs, the 1950s heralded a new Antarctic age – that of science and exploration. Prompted in part by two significant international scientific collaborations, the International Polar Year (1932–3) and International Geophysical Year (1957–8), the age of science and exploration paved the way for the Antarctica we know today – a continent of peace and science.

Top right and above: Sturdy Hagglunds and inflatable rubber boats, as well as tractors, quad bikes and snowmobiles, are the workhorses for scientists living in Antarctica, conducting research in atmospheric studies, plant and animal biology, glaciology, paleontology, medicine, human impact and climate change.

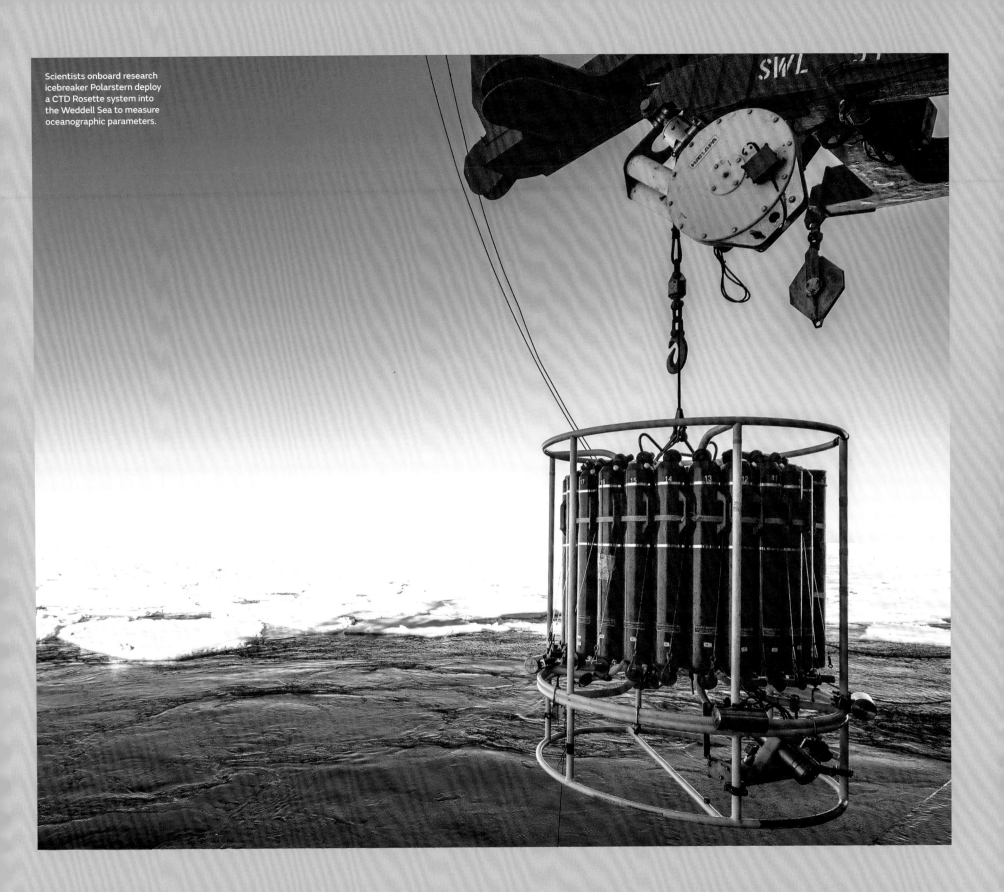

Scientists onboard research icebreaker Polarstern deploy a CTD Rosette system into the Weddell Sea to measure oceanographic parameters.

CHAPTER 5

SHARING ANTARCTICA

No one holds an Antarctic passport. Antarctica has no citizens, no permanent human residents and no Indigenous population. It has no currency and no official language, and yet every year thousands of people flock to this icebound land, some fondly calling it a second home.

Previous: Gentoo penguins share
Argentina's Brown Station, Paradise
Harbour, Antarctic Peninsula.
This page: Focusing on a chinstrap
penguin (*Pygoscelis antarcticus*).
Penguins are generally curious –
this one may have approached the
photographer and is closer than
guidelines recommended. It is not
known whether tourism disturbs
their breeding patterns.

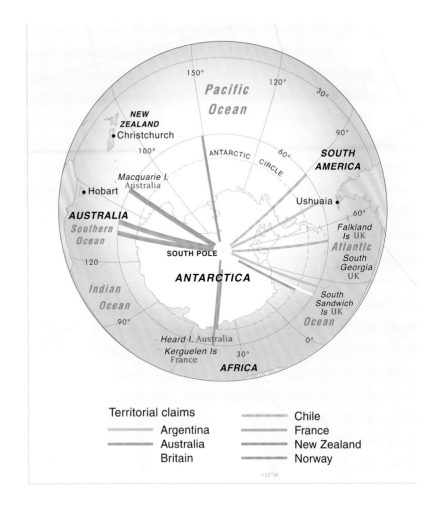

Territorial claims

Argentina		Chile
Australia		France
Britain		New Zealand
		Norway

HUMAN MANAGEMENT

EACH YEAR, THOUSANDS of temporary 'Antarcticans' arrive on the continent – scientists, support staff, visitors and dignitaries, who stay from a few days to more than a year. Scattered across the ice there are roughly 70 scientific stations owned by 30 different countries. The average summer population is around 4000, supplemented by an ever-growing contingent of polar tourists. In winter the human population drops to little more than 1000.

How do we manage this international territory? Who owns it, who polices it and who is in charge when there's no government? The answers are, much like Antarctica itself, unlike anywhere else on Earth.

In the 19th century, Antarctica was a lawless frontier. British and Argentine sealers scuffled over fur seal colonies, and Norwegian and British whalers plied the southern seas in search of the leviathans of the deep. With no formal governance, Antarctica and the ocean surrounding it was an unregulated global commons, plundered without restraint. But the question of who owned Antarctica was already a source of international interest.

Great Britain made the first official Antarctic claim in 1908, which covered the Antarctic Peninsula and several subantarctic islands. When other nations heard about this they followed suit, and by 1940 much of Antarctica had been claimed. France, Norway, Chile, Argentina, Australia and New Zealand had made claims to parts of Antarctica divided along lines of longitude meeting at the South Pole, each nation taking a pie-shaped piece.

A bitter feud erupted between Argentina, Great Britain and Chile, whose claims overlapped. Each eager to bolster their claim, they became embroiled in a game of tit for tat which culminated in Great Britain building three bases across the Antarctic Peninsula, establishing the first year-round presence on the continent. $\longrightarrow$

Above: Antarctic claims are divided along lines of longitude. Right: The Penguin Post Office at British base, Port Lockroy, on the Antarctic Peninsula, is the most southerly operational post office in the world.

Nations that had yet to make a claim in Antarctica sought to assert their sovereignty in other ways. In the summer of 1938–39, German planes flew over the Norwegian-claimed Queen Maud Land, dropping small metal swastikas on the ice. The USA and the USSR refused to acknowledge the existing Antarctic claims, and reserved the right to make a claim in the future.

Tensions escalated during World War II. Rumours that German U-boats were patrolling the Southern Ocean began circulating amongst Allied forces. In 1941 Germany sent a raiding ship into Antarctic waters in search of whale oil. They seized control of eight Norwegian whaling ships and hijacked over 20,000t of whale oil to be used in explosives.

The presence of war ships in Antarctic waters was a warning sign for governments across the world. As a vast and mostly undiscovered territory, Antarctica represented a unique threat during times of war, and both the USA and the USSR harboured fears that Antarctica could be used for military activities. The following decade, as the Cold War deepened, Antarctica loomed as a source of potential conflict in an increasingly unstable world order.

The International Geophysical Year (IGY) program of 1957–8 came at the height of the Cold War.

Scientists proposed a year of collaborative scientific endeavour to boost human understanding of Earth as a global system, including glaciology, oceanography, geology, geomagnetism, polar auroras and meteorology. As one of the least understood areas on the planet, Antarctica was a strong focus of the program.

Despite ongoing political tensions, 12 countries accepted the invitation to participate in the Antarctic program – Australia, Argentina, Belgium, Chile, France, Japan, New Zealand, Norway, South Africa, Great Britain, the USA, and the USSR. Working across 54 stations, these nations enjoyed a period of harmony, supporting one another logistically, sharing scientific data and making widespread observations that helped shape the development of the physical sciences for decades to come.
Eager to maintain this positive momentum, the Scientific Committee on Antarctic Research (SCAR) was established in February 1958, to coordinate ongoing scientific work in Antarctica. $\longrightarrow$

Left: Casey's station buildings are typical of Antarctica's purely functional architecture. Above: This geodesic dome, 50m wide by 16m high, housed the Amundsen–Scott South Pole station from 1975–2003 until it was replaced by a large bunker-style building.
Below: Flags from the original 12 treaty nations fly at the Amundsen–Scott base (USA) at the South Pole.

UN Secretary General, Ban Ki-moon, speaks to reporters during his visit to Antarctica in November 2007. He said "All this may be gone, and not in the distant future, unless we act, together, now. Antarctica is on the verge of a catastrophe – for the world."

In June 1958 the USA invited the 12 nations involved in the Antarctic IGY program to discuss a system of Antarctic governance.

Representatives from all 12 nations assembled in New York. The USA and the USSR put their Cold War tensions aside, and Great Britain, Argentina and Chile put their conflicts on ice. After lengthy negotiations spanning 60 meetings, they drafted the Antarctic Treaty, enshrining in law the values of scientific freedom and warm cooperation that had flourished during the IGY.

This landmark legal document, signed on 1 December 1959 ushered in a "Cold Peace", as Australian historian Tom Griffiths wrote, "established during a brief interglacial in the Cold War". Entered into force in 1961, the treaty outlines a framework for the collective governance of the world's last great wilderness, including all of the land and ice shelves south of 60°S latitude.

Under the treaty, Antarctica can be used for peaceful purposes only. Military activities are prohibited, although military personnel are permitted to support scientific research. A culture of trust and transparency is paramount – all member nations are encouraged to make impromptu site inspections and conduct aerial surveillance of foreign research bases. Existing territorial claims are held in abeyance for the duration of the treaty and no new claims can be made, an approach some have termed 'diplomatic freezing'. Visitors to Antarctica are subject to the laws of their home country, including the international requirements of the Antarctic Treaty, which are enshrined in national legislation.

Since the first Antarctic Treaty Consultative Meeting (ATCM) in 1961, regular meetings between all treaty nations have become an indispensable component of Antarctic governance. Each year representatives meet to discuss issues facing Antarctica, from scientific collaboration to environmental protection, and they are committed to making decisions by consensus. Today, 54 nations are formally committed to the treaty, 29 of which are Consultative Parties with decision-making powers and active science programs in Antarctica.

When the Antarctic Treaty was written, it contained no provisions for activities such as fishing, mining and tourism. Over the past 60 years Antarctic governance arrangements have evolved to encompass several robust international agreements to manage and protect the Antarctic environment, wildlife and human heritage. Together, these fall under the umbrella of the Antarctic Treaty system.

Antarctic fur seals are found mainly on subantarctic islands south of the Antarctic convergence. Their populations have slowly recovered since the 1950s to an estimated one miliion animals. They are still considered a threatened species.

Left: Remains of a former whaling station on black sand beach of Whalers Bay, Deception Island, South Shetland Islands. Below: A raft of gentoo penguins (*Pygoscelis papua*) rests at the surface after diving below for krill.

WILDLIFE CONSERVATION

ANTARCTIC FUR SEALS were hunted almost to extinction in the 1800s. In 1972, in response to a resurgence of commercial interest in Antarctic sealing, the Convention for the Conservation of Antarctic Seals (CCAS) was established. Antarctic sealing has not recommenced since.

In the 1970s commercial fisheries began targeting krill, a small crustacean and a keystone species within the Antarctic ecosystem, in increasing numbers. This prompted the development of the Convention on the Conservation of Antarctic Marine Living Resources (CCAMLR), signed on 20 May 1980.

CCAMLR relates to all marine-living species, except whales, living within the Antarctic Convergence – roughly 10 per cent of the world's oceans. Its purpose is to conserve Antarctic marine life, and ensure any harvesting is sustainable for the whole ecosystem. CCAMLR has also committed to creating a representative network of Marine Protected Areas (MPAs) across the Southern Ocean. Over the past decade CCAMLR has established the South Orkney Islands MPA (2009), and the Ross Sea Region MPA (2016), which encompass approximately 5 per cent of the Southern Ocean. Negotiations continue over three additional proposed MPAs, one in East Antarctica, one in the Weddell Sea and one adjacent to the Antarctic Peninsula.

During the 20th century unsustainable industrial whaling killed more than 2 million whales in the Southern Hemisphere. Most were taken in Antarctic waters. Many whale species were at the edge of extinction when, in 1986, the International Whaling Commission (IWC) implemented a global moratorium on commercial whaling. Japan continued to take minke and fin whales from Antarctic waters under Article VIII of the International Convention for the Regulation of Whaling (ICRW), which allows member nations to issue permits for the purpose of scientific research. In July 2019 Japan withdrew from the ICRW and ceased whaling in the Antarctic.

'THE PARTIES HEREBY DESIGNATE
ANTARCTICA AS A NATURAL RESERVE,
DEVOTED TO PEACE AND SCIENCE'

THE MADRID PROTOCOL, 1991

ENVIRONMENTAL PROTOCOL

THE QUESTION OF mining in Antarctica prompted treaty parties to consider how they might regulate a possible future Antarctic mining industry. They prepared an international convention, which did not enter into force following objections by Australia and France. Instead, the parties drafted the historic Protocol on Environmental Protection to the Antarctic Treaty (also known as the Madrid Protocol). It was signed on 4 October 1991 and entered into force in 1998.

This agreement was a watershed for Antarctic conservation, designating Antarctica as a "natural reserve, devoted to peace and science", with an indefinite ban on mining. Until 2048 the Protocol can only be modified by unanimous agreement of all consultative parties to the Antarctic Treaty. After 2048, any change to the mining ban must include a binding legal regime for Antarctic mineral resource activities.

It would also need to be adopted by a majority of all consultative parties, including three-quarters of the nations who were consultative parties when the Protocol was adopted in 1991. It must also be formally implemented by three-quarters of consultative parties (including all 1991 consultative parties).

While 2048 is sometimes discussed as marking an easing of the conditions around modifying the Madrid Protocol, there is no indication that any nation would be supportive of mining in Antarctica. At the 2016 ATCM treaty parties reaffirmed their commitment to maintaining the ban on mining in Antarctica "as a matter of highest priority".

An Antarctic minke whale swims around Antarctic tourists in zodiacs. Boats must keep a recommended distance from marine life but some animals are curious enough to approach.

Opposite: Tourists outnumber Emperor penguins near Russian icebreaker *Kapitan Khlebnikov*. Right: Adventure tourists camp on the snow at Kerr Point, on Ronge Island, Antarctic Peninsula.

TOURISM

SINCE THE FIRST tourists flew over Antarctica in 1956, interest in experiencing the 'white continent' has steadily increased. Most early Antarctic tourists sailed south in small ice-strengthened ships, primarily to the Antarctic Peninsula, the northernmost part of the continent. Disembarking in small tenders, they cruised around icebergs and visited penguin colonies, historic sites and scientific stations.

Today this style of travel, known as expedition cruising, remains the most popular mode of travel to Antarctica. These voyages may offer adventurous options such as sea kayaking. Smaller numbers travel by yacht or in large, cruise-only vessels, which don't land. For a fortunate few, remote field camps offer luxury accommodation, tours and extended expeditions, including mountaineering and skiing.

It's even possible to join a gruelling Antarctic ultra-marathon over 100km of ice and snow.

Antarctic tourism has grown exponentially in recent years, with more than 55,000 tourists visiting in the summer of 2018–19. This rapid growth has prompted concerns about the potential risks of welcoming so many visitors to such a remote, natural environment. The difficulty of coordinating a rescue in the event of a maritime incident is one such concern. Other considerations include the over-crowding of existing landing sites, expansion into new areas, waste management, and cumulative impacts on sensitive areas, including the introduction of foreign species or diseases. This threat is compounded by climate change, which is opening up new, warmer habitats across the northern regions. $\longrightarrow$

The 100km Antarctic ice marathon is run annually in the lee of Patriot Hills, just south of the Antarctic Peninsula. Athletes face an average windchill temperature of −20°C, the possibility of strong katabatic winds and an average course elevation of 900m.

Under the Environmental Protocol and rules adopted by the ATCM, tourism activities (and all other Antarctic activities) are subject to regulations administered by each treaty party, including requirements for environmental impact assessments, rules about where activities are allowed, and guidelines on how to operate safely and without harming the environment. Environmental concerns are also addressed by the International Association of Antarctica Tour Operators (IAATO) whose members adhere to IAATO procedures and guidelines to ensure safe, environmentally sound travel. These include regulations on biosecurity and guidelines for visiting specific sites and wildlife watching. Looking to the future, the impact of the 2020 global pandemic on ship-based tourism in Antarctica remains uncertain.

THE ANTARCTIC TREATY celebrated its 60th anniversary in December 2019. In the 60 years since its inception the Treaty has adapted to emergent industries and an ever-changing geopolitical landscape. It has successfully protected an entire continent from the ravages of war and resource extraction, and fostered a flourishing culture of collaborative scientific endeavour. The original twelve parties and the consultative parties meet annually, and special meetings may be called when particular topics need to be addressed.

Often celebrated as one of the world's most successful international agreements, the future success of the Antarctic Treaty System relies on the continued goodwill and cooperation of all treaty nations. The qualities of compromise, cooperation and diplomacy have never been more important. The coming decades will present new challenges, as Antarctica faces the combined pressures of an evolving tourism industry, global resource scarcity and climate change.

A superb way to experience the daunting scale of Antarctic scenery – kayakers at Paradise Harbour, Antarctic Peninsula.

CHAPTER 6

CLIMATE CHANGE

Antarctica is among the fastest warming areas on Earth. The spectre of melting ice caps and rising sea levels has sparked a sense of urgency to investigate how a rapidly warming climate will affect humanity. Scientists are looking to Antarctica as both an archive of past climates and sentinel of the changes to come.

Previous: Ice calving from a glacier. Antarctic ice is melting around six times faster than in the 1990s. Left: Sea ice typically reaches its annual maximum extent in September and minimum extent in early March.

RAPID WARMING

WHEN THE ANTARCTIC ice sheet formed about 34 million years ago it heralded the beginning of an ice age that continues today. Since then glaciers have advanced and retreated, and sea levels have risen and fallen through global periods of relative warmth and cooling, called glacial and interglacial periods.

Fluctuations in climate, which normally take place over tens of thousands or millions of years, are spurred on by natural factors, some chaotic, others cyclical. Right now we are living in an interglacial period, a period of relative stability which started about 10,000 years ago. But since the industrial era began in the late 19th century, our climate has become unsettled. While natural climate cycles suggest we should be entering a cooling period, the planet is rapidly heating as greenhouse gases, like carbon dioxide, accumulate in the atmosphere. Human activities, such as burning fossil fuels,

are speeding up the release of greenhouse gases, which are usually slowly released and reabsorbed by natural systems, over millennia.

In 2007 the Intergovernmental Panel on Climate Change (IPCC) declared "warming of the climate system is unequivocal, and since the 1950s, many of the observed changes are unprecedented over decades to millennia". Today, atmospheric carbon dioxide levels are the highest they've been in 3 million years.

Dr Jess Melbourne-Thomas operates a sea ice corer. Ice cores are used not only to study climate history, but also to study organisms such as algae. Ice algae are an important source of food for krill and other zooplankton.

Left: Australian Antarctic Division glaciologist, Dr Tas van Ommen, processes an ice core from Law Dome in East Antarctica.
Below: Extracting a core from the drill at Law Dome, 100km southeast of Australia's Casey station.

ICE CORES –
A WINDOW TO THE PAST

DEEP BENEATH THE surface of Antarctica's ice sheet are millions of tiny time capsules – bubbles of ancient atmosphere trapped within a frozen matrix. These pockets of air allow scientists to peer into the past to a time when the ice fell as snow. Researchers have drilled up to 3km into some of the oldest ice on Earth, extracting cylindrical sections (cores) of ice up to 6m long and about 13cm in diameter. The oldest of these cores goes back 800,000 years.

By measuring the concentration of gases like carbon dioxide and methane, and aerosols like pollen and ash, paleo-climatologists are able to map out the deep history of our atmosphere and climate,

from warm and cold spells to volcanic eruptions, providing a historical context for the changes we are witnessing today. Data from ice cores shows that the rate of rise in greenhouse gases since the Industrial Revolution is almost certainly unprecedented over the past 800,000 years.

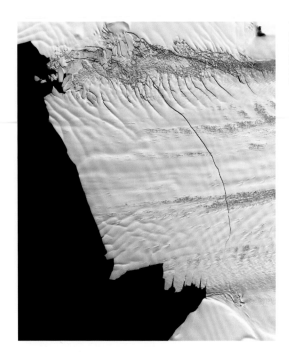

This NASA Terra spacecraft image from 2011 shows a massive crack across the Pine Island Glacier, a major ice stream that drains the West Antarctic Ice Sheet.

THE ANTARCTIC ICE SHEET

GLACIOLOGISTS OFTEN REFER to the Antarctic ice sheet as the 'sleeping giant'. For much of human history this vast mass of ice, with the potential to submerge continental coastlines and raise global sea levels by 57m, has lain dormant. But now the giant is stirring.

Antarctic ice is melting around six times faster than in the 1990s, and the rate of global sea level rise is accelerating every year. Of the 252 gigatons of ice breaking off Antarctica annually, the vast majority comes from the West Antarctic Ice Sheet (WAIS). Holding ten per cent of the continent's ice, much of the WAIS is grounded on bedrock below sea level, which slopes gradually inland. This quirk of topography exposes glaciers to seawater eroding them from below, making them highly vulnerable to melting.

Most of these low-lying glaciers are stabilised by ice shelves, formed when glaciers flow over the ocean in floating sheets. Ice shelves shore up the immense glaciers flowing off the continent behind them. When the Antarctic Peninsula's Larsen B Ice Shelf disintegrated in 2002, the glaciers that had fed the ice shelf began to flow around six times faster. Ice shelves are one of the most vulnerable features of the Antarctic icescape, thinning dramatically as warm seawater laps at their undersides and peels them off their foundations on the sea floor. $\longrightarrow$

Crevasse-riven glacier on the remnant Larsen B Ice Shelf.

"THE PRIMARY DRIVER OF MELTING IN ANTARCTICA IS WHERE THE ICE MEETS THE OCEAN."

DR TAS VAN OMMEN

NASA's Operation IceBridge conducted aerial surveys of polar regions for 11 years until 2019 to record changes in polar ice. Data on the thickness and shape of snow and ice, as well as the topography of the land and ocean floor, has allowed scientists to determine that the West Antarctic Ice Sheet, see here in 2017, may be in irreversible decline.

This forces the grounding line – the point where ice detaches from the land and begins to float – inland, exposing ever more ice to the persistent pulse of the sea.

Until very recently, scientists thought the East Antarctic Ice Sheet (EAIS), a colossus containing enough ice to raise sea levels by more than 50m, was perched high on dry land, invulnerable to rising temperatures. But new research reveals even the coldest place on Earth is not immune to a warming climate.

In the International Polar Year in 2008 an international consortium of researchers formed ICECAP (International Collaboration for Exploration of the Cryosphere through Aerogeophysical Profiling) with the aim of gaining a clearer picture of the contours of the earth beneath the smooth ice cap. Their aerial surveys charted two deep basins plunging far below sea level beneath the EAIS. East Antarctica, it seemed, had more in common with West Antarctica than originally thought. Australian Antarctic Division glaciologist Dr Tas van Ommen, a lead investigator on the project, says "there could be three to six metres of sea level rise coming from East Antarctica with only modest retreat of the ice there". A one metre rise in sea level would displace up to 230 million people.

These findings, coupled with mounting evidence that parts of the EAIS have melted during warm spells in the recent past, are revolutionising our understanding of how this icy behemoth might react to ongoing climate change. To truly understand the forces at play, however, researchers are looking to the sea. "That's where all the activity is," says Dr van Ommen. "The primary driver of melting [in Antarctica] is where the ocean meets the ice."

The West Ice Shelf lies on the King Leopold and Queen Astrid Coast, Princess Elizabeth Land. The ice shelf, one of the largest in East Antarctica, extends 350km between Barrier Bay in the west and Posadowsky Bay in the east.

CARBON IN
THE DEPTHS

THE OCEAN, ATMOSPHERE and cryosphere (ice) are engaged in an intimate and age-old dance – a constant exchange of particles and gases through evaporation, condensation and absorption, catalysed by currents, winds, snow and rain. Together, these forces drive global climate and weather patterns.

One way the ocean moderates the global climate is through the absorption of heat and carbon dioxide. The ocean is a very effective heat sink, and over the past 50 years it has absorbed 93 per cent of the global increase in warmth caused by human activity – the majority of this in the Southern Ocean. The oceans have also absorbed up to 30 per cent of the excess carbon dioxide generated by human activity, much of this in cold polar waters, which can hold more dissolved gas than warmer water.

Southern Ocean circulation patterns mean much of this excess heat and carbon dioxide has been transported into deep ocean currents, where it may not resurface for decades or even millennia. The ocean is "doing us a great service" by removing this excess heat and carbon dioxide from the atmosphere, says Nathan Bindoff, a physical oceanographer at the Institute for Marine and Antarctic Studies (IMAS), "but it comes at a cost." A warmer ocean causes more rapid melting of ice shelves. It also expands, taking up more space and exacerbating sea level rise. When carbon dioxide is absorbed by the ocean, the water becomes more acidic, making it more difficult for some marine species to build and maintain protective shells.

Since the year 2000 an array of 4000 roving robots called Argo floats has roamed the oceans, providing oceanographers with an unprecedented insight into the changing properties of the sea. "These represent a profound shift in the way we study the oceans," says Dr Steve Rintoul, an oceanographer with the CSIRO. "We can now measure the ocean globally in a way that was never possible before."

Antarctic researchers are closely monitoring shifting patterns in the Southern Ocean, which could have serious ramifications for Antarctic ice and ecosystems, and the global climate. "The Southern Ocean has a thirty-to-forty-year response time," says Bindoff. "What we're looking at in the Southern Ocean is the effect of what we were doing thirty to forty years ago. We're waiting for it to catch up."

Dr Steve Rintoul, CSIRO oceanographer holds an ARGO float, used to measure ocean currents.

Top left: A shy albatross (*Thalassarche cauta*) swoops low over a gale-driven Southern Ocean. Left: King penguins (*Aptenodytes patagonicus*) commonly dive to 100m, and have been recorded at more than 300m, in the Southern Ocean. They feed at the Antarctic Convergence and breed on subantarctic islands. Future ocean warming may drive their hunting grounds too far from favoured breeding sites.

The ice-strengthened hull of Russian-registered vessel *Spirit of Enderby* makes its way through sea ice on a cruise to Mawson's Hut.

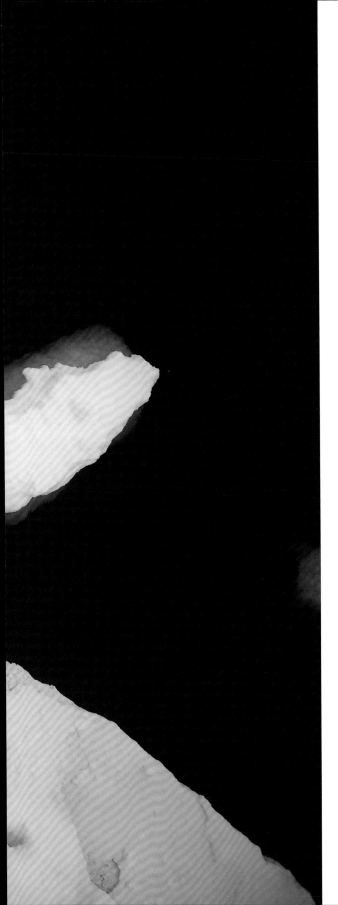

SEA ICE

IN ADDITION TO glacier ice flowing off the land, Antarctica is surrounded by fields of sea ice, or frozen ocean, which expand and contract seasonally. For four decades, from the 1970s, as the oceans warmed and ice shelves melted, this fringe of sea ice was inexplicably growing each winter. It reached a record high in 2014, covering an area more than twice the size of Australia. Before researchers could explain this astonishing growth, Antarctic sea ice cover plummeted. By 2017 it was the lowest it had been in 40 years.

It is still unclear whether this contraction is the beginning of long-term decline or simply a seasonal variation, possibly caused by changes in the ozone layer or weather patterns like El Niño. Scientists are watching closely to see what happens next.

Sea ice loss has many impacts on the Antarctic ecosystem and the global climate. Its white surface reflects as much as 90 per cent of the Sun's heat energy back to space, while the ocean reflects only about 5 per cent, absorbing the rest. Sea ice is also a critical habitat for ocean-dwelling organisms including phytoplankton and krill, cornerstones of the Antarctic ecosystem.

Left: Releasing krill back into ocean. Below: Sorting the krill catch. Krill is extensively studied due to its critical importance in the Antartic foodchain.

Hooker's sea lion (*Phocarctos hookeri*), found on New Zealand's sub-Antarctic Auckland Islands, one of the rare species facing uncertain threats from climate change.

ANTARCTIC ECOSYSTEMS

THE WEST ANTARCTIC Peninsula, a string of mountains and volcanoes curving towards South America, is one of the most rapidly warming areas on Earth. Average air temperatures here have increased by 3°C in 50 years, and the adjacent ocean is now ice free for 90 days longer in summer than it was in 1979.

The Antarctic Peninsula is an oasis for Antarctic wildlife, including krill, a keystone species within the Antarctic ecosystem. These small crustaceans are the primary food source for many fish, seabirds, penguins, seals and whales. But reduced sea ice cover and warming oceans appear to be impacting krill numbers.

"Sea ice is critical for krill," says Rob King, a krill biologist with the Australian Antarctic Division and biology lead in the design of Australia's new icebreaker, RSV *Nuyina*. "It's an important larval habitat. It offers protection from predators and provides a surface on which their food, phytoplankton, can grow."

In a small lab at the Australian Antarctic Division in Tasmania, King and other Australian scientists are studying how krill populations may respond to a warmer, more acidic ocean. Their research suggests that while adult krill aren't particularly vulnerable, their eggs and larvae are. If we continue with business as usual, King says, there will be a 50 per cent reduction in the number of krill eggs hatching by the year 2100, with enormous effects on the Antarctic food chain.

While krill are important sentinels, they are only one small part of the extraordinarily complex Antarctic ecosystem. Dr Andrew Constable, a marine ecologist with the Australian Antarctic Division, emphasises the importance of continued research. "We have a good understanding of what is in the ecosystem but not the dynamics of it. $\longrightarrow$

Left: If the Antarctic ice sheet melted, global sea levels would rise by 57m. Opposite: Southern giant petrels bother a leopard seal. Climate change is starting to impact the behaviour of the leopard seal, a solitary apex predator that normally breeds and feeds on pack ice.

Antarctic marine ecosystems are undergoing unprecedented change and we don't know quite how that's going to transpire." The AAD's new icebreaker, equipped with facilities for ship-based research, promises to make an important contribution to critical Australian-led research, and create a deeper understanding of Antarctic ecosystems.

Perhaps the greatest frontier faced by climate researchers today is the persistent disconnect between scientific research and global decision-making. Dr Jess Melbourne-Thomas is an Antarctic, marine and climate change scientist and recipient of the 2020 Tasmanian Australian of the Year Award. She is one of many researchers who have recognised a pressing need for better transmission of knowledge between scientists, governments, industries and communities. "Many of the problems we have now are very urgent, and we need more efficient ways to connect science to policy." In her role as

a Transdisciplinary Researcher and Knowledge Broker for the CSIRO, Dr Melbourne-Thomas is forging connections to improve these lines of communication and understanding.

Antarctica is at the forefront of a rapidly changing climate. As the Earth's system wavers between accommodation, correction and collapse, it contains clues to help us understand how the climate works and predict what may happen next. Technological advancements are allowing researchers to probe the depths of Antarctic ice and the breadth of the Southern Ocean like never before, unearthing records of past climates and portents of future sea level rise. These clues are helping them anticipate future changes with increasing confidence, so we can prevent and prepare in the best ways possible.

BIBLIOGRAPHY

The following resources were used to research *Antarctica*.
This is not an exhaustive list but includes many interesting sources for Antarctic enthusiasts.

BOOKS

Amundsen, Roald. *The South Pole*. London, John Murray, 1912.

Benedict Allan (Ed). *The Faber Book of Exploration*. London, Faber, 2004

Boothe, Joan N. *The Storied Ice: Exploration, Discovery, and Adventure in Antarctica's Peninsula Region*. Regent Press, 2011

Brady, Anne-Marie. *The Emerging Politics of Antarctica. Routledge*, 2012

Carson, Rachel. *The Sea Around Us*. New York, Oxford University Press, 1951

Crane, D. *Scott of the Antarctic: A Life of Courage, and Tragedy in the Extreme South*. London: HarperCollins, 2005

Day, David. *Antarctica: A Biography*. Sydney, Random House Australia, 2013

Dodds, Hemmings, Roberts (Eds) *Handbook on the Politics of Antarctica*. Edward Elgar Publishing, 2017

Fitzsimons, Peter. *Mawson*. Sydney, William Heinemann Australia, 2014

Franzen, Jonathan. *The End of the End of the Earth*. London, HarperCollins Publishers, 2018.

Garner, Helen. *Regions of Thick-Ribbed Ice*. Melbourne, Black Inc, 2015

Grann, David. *The White Darkness*. New York, Doubleday, 2018.

Griffiths, Tom. *Slicing the Silence: Voyaging to Antarctica*. Harvard University Press, 2009

Howkins, Adrian. *The polar regions*. Wiley, 2015

Land, Barbara. *The New Explorers – Women in Antarctica*. New York, Dodd, Mead, 1981

Liggett, Storey, Cook, Meduna (Eds). *Exploring the Last Continent*. Springer 2015

Mawson, Douglas. *The Home of the Blizzard*. South Australia, Wakefield Press, 1996

McCann, Joy. *Wild Sea: A History of the Southern Ocean*. Sydney, NewSouth Publishing, 2018

Pearson, Michael. *Great Southern Land: The Maritime Exploration of Terra Australis*, Canberra, Dept. of the Environment and Heritage, 2005

Ruddiman, William F. *Earth's Climate, Past and Future*. Second Edition. W. H. Freeman and Company, 2008

Shirihai, Hadoram. *A Complete Guide to Antarctic Wildlife*. London, Bloomsbury, 2019

Smith, Thomas, W. *A Narrative of The Life, Travels and Sufferings of Thomas W. Smith*. Gale, 2012

JOURNALS

Australian Antarctic Magazine
Australian Humanities Review
Canadian Journal of Zoology
Climactic Change
Earth-Science Reviews
EOS
Geophysical Research Letters
Gondwana Research
Journal of Climate
Journal of Crustacean Biology
Journal of Glaciology
Journal of Physical Oceanography
Marine Ecology Progress Series
Nature
Nature Communications
Nature Geoscience
Oceanography
Polar Biology
Procedia Social and Behavioral Sciences
Science
Scientific Reports
The Linnean

WEBSITES

acecrc.org.au (Antarctic Climate & Ecosystems Cooperative Research Centre)
anareclub.org
antarctica.gov.au
antarcticglaciers.org
asoc.org (Antarctic and Southern Ocean Coalition)
ats.aq (Secretariat of the Antarctic Treaty)
australiangeographic.com
australianmuseum.net.au
bas.ac.uk (British Antarctic Survey)
biologicaldiversity.org
carbonbrief.org (UK)
ccamlr.org (Commission for the Conservation of Antarctic Marine Living Resources)
climate.gov (National Oceanic and Atmospheric Administration
climate.nasa.gov (USA)
climatecentral.org

discoveringantarctica.org.uk
iaato.org (International Association of Antarctica Tour Operators)
iced.ac.uk (Integrating Climate and Ecosystem Dynamics)
igsoc.org (International glaciological society, UK)
imas.utas.edu.au (Institute for Marine & Antarctic Studies)
imbie.org (Ice sheet mass balance inter-comparison exercise)
imos.org.au/facilities/argo (Integrated Marine Observing System)
ipcc.ch (Intergovernmental Panel on Climate Change)
iucnredlist.org
iwc.int (International Whaling Commission)
livescience.com
nationalgeographic.com
nature.com
nerc.ukri.org (National Environment Research Council, UK)
newscientist.com
nisdc.org (National Snow and Ice Data Center, USA)
nodc.noaa.gov/woce (World Ocean Circulation Experiment)
nsf.gov (National Science Foundation, USA)
oceanservice.noaa.gov
scar.org (Scientific Committee on Antarctic Research)
sciencedaily.com
sciencedirect.com
sciencex.com
scientificamerican.com
smithsonianmag.com
soccom.princeton.edu (The Southern Ocean Carbon and Climate Observations and Modeling)
soos.aq (Southern Ocean Observing System)
spri.cam.ac.uk (Scott Polar Research Institute)
tos.org (The Oceanography Society)
ukaht.org (UK Antarctic Heritage Trust)
usgs.gov (United States Geological Survey)
usjgofs.whoi.edu (US Joint Global Ocean Flux Study)
wcrp-climate.org (World Climate Research Program)

INDEX

Cover image by Justin Gilligan.
Light and ice converge along the
coast of Buckle Island, part of
the Balleny Islands, a series of
uninhabited Antarctic islands
claimed by New Zealand. Their
volcanic slopes are decorated
by glaciers that fall dramatically
into an icy Southern Ocean.

PHOTO CREDITS

All images are protected by copyright and have been reproduced with permission.
Abbreviations used: TR: top right, TL: top left, L: left; R: right B: bottom, BR: bottom right, BL: bottom left

COVER Justin Gilligan (see opposite); 2 Getty/Posnov; 4 Justin Gilligan; 8 Getty/Mike Hill; 10 NASA Image Library; 12 Australian Antarctic Division (AAD)/Patrick James; 14 AAD/Chris Crerar; 15 Mitchell Library, State Library of NSW (SLNSW); 16 Australian Geographic (AG)/Chrissie Goldrick; 17 AAD/Reed Burgette; 18 AAD/Victoria Heinrich; 19 AG/Glen Foote; 20 AAD/Robyn Mundy; 21 AAD/Ian Phillips; 22-23 AAD/Christopher Wilson; 24 CSIRO/Peter Kimball; 25TR AAD/Paul Hanlon; 25B AAD; 25TL AAD/Greg Stone; 26–27 AAD/Justin Chambers; 28 AG; 29 AAD/Kirsten le Mar; 31 AAD/Glenn Johnstone; 32 Alamy/Colin Monteath/Minden Pictures; 33 Alamy/Colin Harris; 34–35 NASA Image Library; 36 Getty/Posnov; 38 Shutterstock/Armin Rose; 39TL Shutterstock/Armin Rose; 39TR Alamy/Justin Hofman; 40–41 Alamy/Michel Roggo; 42 Getty/Yann Arthus-Bertrand; 43 Getty/John Borthwick; 44–45 Getty/Yann Arthus-Bertrand; 46 NASA Image Library; 49 Getty/Xavier Hoenner Photography; 50 Shutterstock/Jo Crebbin; 52 AG/Chrissie Goldrick; 53TL AAD/Charles Millen; 53TR Shutterstock/Szakharov; 53B AG/Jasmine Poole; 54–55 Shutterstock/Mariusz Potocki; 57 Shutterstock/Vlad Silver; 58TL iStock/pilipenkoD; 58BR Shutterstock/Incredible arctic; 58BL iStock/mzphoto11; 58TR iStock/Goddard_Photography; 61 AAD/Patti Virtue; 62T Shutterstock/Chase Dekker; 62BL Alamy/Sea Tops; 62BR Shutterstock/Stephen Lew; 63 Shutterstock/MZPHOTO.CZ; 64 AAD/Brett Wilks; 65L Alamy/Colin Harris/era-images; 65R iStock/mike_matas; 66 AG/Jasmine Poole; 69 The University of Melbourne Library Collection; 70–71 Alamy/The Picture Art Collection; 72–73 Justin Gilligan; 74 National Library of Australia (NLA)/Frank Hurley; 76 NLA/Frank Hurley; 77L Getty/Bettmann; 77TL AG/Jasmine Poole; 77BR SLNSW/Percy Gray; 78L SLNSW/Frank Hurley; 78R AG/Mike Rossi; 79 NLA/Frank Hurley; 80 Antarctica New Zealand Pictorial Collection/Dr Graeme Midwinter; 81R Museums Victoria/George Rayner; 81L Antarctica New Zealand Pictorial Collection/John Claydon; 83TL Shutterstock/Armin Rose; 83TR AAD/Christopher R. Clarke; 83B Getty/Mint Images; 83 iStock/Tenedos; 85 Justin Walker; 86 Shutterstock/Yongyut Kumsri; 87 Shutterstock/Travel Media Productions; 88–89 AAD/Barry Becker; 90TL AAD/Doug McVeigh; 90TR Getty/Galen Rowell; 90B Shutterstock/mhelm4; 91 Getty/AFP/Rodrigo Arangua; 92 AG/Chrissie Goldrick; 93L Shutterstock/TasfotoNL; 93R Getty/David Merron Photography; 94–95 iStock/Andrew Peacock; 96 AAD; 97 AG/Mike Rossi; 98 Getty/Alejandro Jinich Diamant; 99 Justin Walker; 100 worldmarathons.com; 101 Justin Walker; 102 Shutterstock/Bernhard Staehli; 104 Justin Gilligan; 106 CSIRO/Brian Walpole; 107L AAD/Joel Pedro; 107R AAD/Gordon Tait; 108 NASA Image Library; 109 Shutterstock/Armin Rose; 110–111 Justin Gilligan; 112 NASA Image Library; 113 AAD/Kerry Steinberner; 115TL iStock/Mickrick; 115BL Alamy/Danita Delimont Creative; 115R Alamy/Bruce Miller; 116 AG/Mike Rossi; 118TL AAD/Rob King; 118TR AAD/Brett Free; 118B Justin Gilligan; 120 AAD/Charles Millen; 121 AAD/Peter Layt; 122 AAD/Chris Wilson; 126 Justin Gilligan.

Australian Geographic

Published in 2020 by Australian Geographic
Level 7, 54 Park Street, Sydney NSW 2000
Telephone 02 9136 7214
Email editorial@ausgeo.com.au
Website australiangeographic.com.au

Editor Katrina O'Brien
Sub-editor Rebecca Cotton
Creative director Mike Ellott
Senior Designer Harmony Southern
Cartography Will Pringle

Australian Geographic Editor-in-chief Chrissie Goldrick
Australian Geographic Managing director Jo Runciman
ISBN 978-1-925847-86-4

Printed and bound in China by C&C Offset Printing Co. Ltd.

The **Australian Geographic Society** was established to encourage the spirit of discovery and adventure, and to foster love for our natural heritage. The Society and the Australian Geographic journal sponsor scientific research and conservation, and a portion of the profits from our published products goes back into the Society. Become a member today by subscribing to the Australian Geographic journal.

Call now: 1300 555 176 (within Australia)

Acknowledgements
Nina Gallo would like to thank all the scientists and support staff who contribute to the Australian Antarctic Program, with special thanks to Nathan Bindoff, Dr Jaimie Cleeland, Dr Andrew Constable, Dr Rachel Hawker, Rob King, Delphine Lannuzel, Dr Jess Melbourne-Thomas, Dr Steve Rintoul, Mic Rofe, Dr Aleks Terauds, Dr Tas van Ommen, Dr Tessa Vance.
She would also like to thank:
- All the incredible people I've had the privilege of working with in Antarctica.
- My friends, family and peers for sitting through hours of talk about Antarctica and reading my words with a critical eye. This would have been less without you.
Finally, she would like to thank all the nations, organisations and individuals working tirelessly to ensure the continued success of the collective governance, study and protection of Antarctica.

Thanks to all the photographers whose work is included here. We would especially like to thank Justin Gilligan and Justin Walker.
Australian Geographic would also like to acknowledge the assistance of the Australian Antarctic Division and CSIRO.

A catalogue record for this book is available from the National Library of Australia

FSC
www.fsc.org
MIX
Paper from responsible sources
FSC® C008047